MATHEMATICS OF GENOME ANALYSIS

The massive research effort known as the Human Genome Project is an attempt to record the sequence of the three billion nucleotides that make up the human genome and to identify individual genes within this sequence. Although the basic effort is of course a biological one, the description and classification of sequences also naturally lend themselves to mathematical and statistical modeling.

This short textbook on the mathematics of genome analysis presents a brief description of several ways in which mathematics and statistics are being used in genome analysis and sequencing. It will be of interest not only to students but also to professional mathematicians curious about the subject.

Jerome K. Percus is Professor of Physics and Mathematics at the Courant Institute of Mathematical Sciences and Department of Physics at New York University, where he has taught since 1958. He has held visiting positions at Middlesex Hospital Medical School in London, Columbia University, Rutgers University, Princeton University, Rockefeller University, Yukawa Institute in Kyoto, Tokyo University, Norwegian Institute of Technology, Max Planck Institute in Tubingen, Catholic University in Rio de Janeiro, Ecole Polytechnique in Lausanne, Soviet Academy of Sciences in Moscow, Leningrad, Kiev, and Lvov, University of Paris, Nankai University, and Tsinghua University in China. He has received the Pregel (New York Academy of Science), Pattern Recognition Society, and Hildebrand (American Chemical Society) Chemical Physics awards.

Cambridge Studies in Mathematical Biology

Editors

C. CANNINGS
University of Sheffield, UK

F. C. HOPPENSTEADT
Arizona State University, Tempe, USA

L. A. SEGEL
Weizmann Institute of Science, Rehovot, Israel

MATHEMATICS OF GENOME ANALYSIS

JEROME K. PERCUS
New York University

CAMBRIDGE
UNIVERSITY PRESS

PUBLISHED BY THE PRESS SYNDICATE OF THE UNIVERSITY OF CAMBRIDGE
The Pitt Building, Trumpington Street, Cambridge, United Kingdom

CAMBRIDGE UNIVERSITY PRESS
The Edinburgh Building, Cambridge CB2 2RU, UK
40 West 20th Street, New York, NY 10011-4211, USA
10 Stamford Road, Oakleigh, VIC 3166, Australia
Ruiz de Alarcón 13, 28014 Madrid, Spain
Dock House, The Waterfront, Cape Town 8001, South Africa

http://www.cambridge.org

First published 2002

Printed in the United Kingdom at the University Press, Cambridge

Typeface Times Roman 10.25/13 pt.   *System* LaTeX 2_ε [TB]

A catalog record for this book is available from the British Library.

Library of Congress Cataloging in Publication Data
Percus, Jerome K. (Jerome Kenneth)
Mathematics of genome analysis / Jerome K. Percus.
p.   cm. – (Cambridge studies in mathematical biology ; 17)
Includes bibliographical references and index.
ISBN 0-521-58517-1 – ISBN 0-521-58526-0 (pb.)
1. Genetics – Mathematical models.   2. Genetics – Statistical methods.   3. Gene mapping –
Mathematical models.   4. Gene mapping – Statistical methods.   I. Title.   II. Series.
QH438.4.M3 P47   2001
572.8'6'0151 – dc21   2001035087

ISBN 0 521 58517 1 hardback
ISBN 0 521 58526 0 paperback

Contents

Preface

"What is life?" is a perennial question of transcendent importance that can be addressed at a bewildering set of levels. A quantitative scientist, met with such a question, will tend to adopt a reductionist attitude and first seek the discernible units of the system under study. These are, to be sure, molecular, but it has become clear only in recent decades that the "founder" molecules share the primary structure of linear sequences – in accord with a temporal sequence of construction – subsequent to which chemical binding as well as excision can both embed the sequence meaningfully in real space and create a much larger population of molecular species. At the level of sequences, this characterization is, not surprisingly, an oversimplification because, overwhelmingly, the construction process of a life form proceeds via the linear sequences of DNA, then of RNA, then of protein, on the way to an explosion of types of molecular species. The founder population of *this* subsequence is certainly DNA, which is the principal focus of our study, but not – from an informational viewpoint – to the exclusion of the proteins that serve as ubiquitous enzymes, as well as messengers and structural elements; the fascinating story of RNA will be referred to only obliquely.

That the molecules we have to deal with are fundamentally describable as ordered linear sequences is a great blessing to the quantitatively attuned. Methods of statistical physics are particularly adept at treating such entities, and information science – the implied context of the bulk of our considerations – is also most comfortable with these objects. This hardly translates to triviality, as a moment's reflection on the structure of human language will make evident.

In the hyperactive field that "genomics" has become, the focus evolves very rapidly, and "traditional" may refer to activities two or three years old. I first presented the bulk of this material to a highly heterogeneous class in 1993, again with modification in 1996, and once more, further modified, in 1999. The aim was to set forth the mathematical framework in which the burgeoning activity takes place, and, although hardly impervious to the passage of time,

ix

this approach imparts a certain amount of stability to an intrinsically unstable divergent structure. I do, of course, take advantage of this nominal stability, leaving it to the reader to consult the numerous technical journals, as well as (with due caution) the increasing flood of semitechnical articles that document the important emerging facets of the current genomics field.

It is a pleasure to acknowledge the help of Connie Engle and Daisy Calderon-Mojar in converting a largely illegible manuscript to, it is hoped, readable form, of Professor Ora Percus for insisting on reducing the non sequiturs with which the original manuscript abounded, and of numerous students who not only maintained intelligent faces, but also gently pointed out instances of confusion in the original lectures.

1

Decomposing DNA

1.1. DNA Sequences

The realization that the genetic blueprint of a living organism is recorded in its DNA molecules developed over more than a century – slowly on the scale of the lifetime of the individual, but instantaneously on the scale of societal development. Divining the fashion in which this information is used by the organism is an enormous challenge that promises to dominate the life sciences for the foreseeable future. A crucial preliminary is, of course, that of actually compiling the sequence that defines the DNA of a given organism, and a fair amount of effort is devoted here to examples of how this has been and is being accomplished. We focus on nuclear DNA, ignoring the miniscule mitochondrial DNA.

To start, let us introduce the major actor in the current show of life, the DNA chain, a very long polymer with a high degree of commonality – 99.8%, to within rearrangement of sections – among members of a given species [see Alberts et al. (1989) for an encyclopedic account of the biology, Cooper (1992) for a brief version, Miura (1986), and Gindikin (1992) for brief mathematical overviews]. The backbone of the DNA polymer is an alternating chain of *phosphate* (PO_4) and sugar (S) groups. The sugar is *deoxyribose* (an unmarked vertex in its diagrammatic representation always

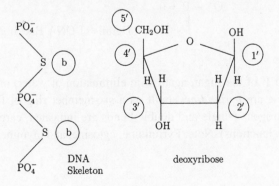

1

signifies a carbon atom) with standard identification of the five carbons
as shown. Successive sugars are joined by a phosphate group (phosphoric
acid, H_3PO_4, in which we can imagine that two hydrogens have combined
with 3′ and 5′OHs groups of the sugar, with the elimination of water,
whereas one hydrogen has disappeared to create a negative ion); the whole
chain then has a characteristic 5′–3′ orientation (left to right in typical
diagrams, corresponding to the direction of "reading," also upstream to downstream). However, the crucial components are the side chains or bases

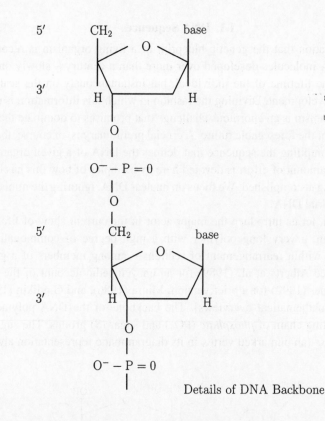

Details of DNA Backbone

(attached to 1′ of the sugar, again with elimination of water) of four types.
Two of these are *pyrimidines*, built on a six-member ring of four carbons
and two nitrogens (single and double bonds are indicated, carbons are implicit at line junctions). Note: Pyrimidine, cytosine, and thymine all have the
letter *y*.

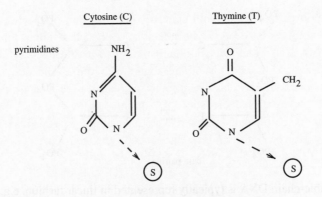

Two are the more bulky *purines*, built on joined five- and six-member rings (adenine, with empirical formula $H_5C_5N_5$, used to have the threatening name pentahydrogen cyanide, of possible evolutionary significance).

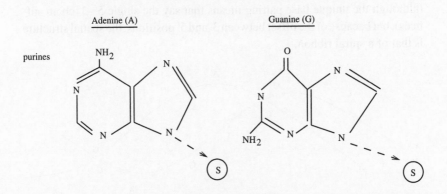

DNA chains are normally present as pairs, in the famous Watson–Crick *double-helix* conformation, enhancing their mechanical integrity. The two strands are bound through pairs of bases, pyrimidines to purines, by means of *hydrogen bonds* (......), and chemical fitting requires that A must pair with T, G with C; thus each chain uniquely determines its partner. The DNA "alphabet" consists of only the four letters A, T, G, and C, but the full text is very long indeed, some 3×10^9 base pairs in the human. Roughly 3% of *our* DNA four-letter information is allocated to genes, "words" that translate into the proteins that, among other activities, create the enzymatic machinery that drives biochemistry, as well as instructional elements, the rest having unknown – perhaps mechanical – function.

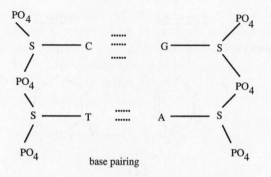

base pairing

Double-chain DNA is typically represented in linear fashion, e.g.,

$$5' - A - C - G - T - G - A - C - 3'$$
$$\vdots \quad \vdots \quad \vdots \quad \vdots \quad \vdots \quad \vdots \quad \vdots$$
$$3' - T - G - C - A - C - T - G - 5'$$

(although the unique base pairing means that say the single 5'–3' chain suffices), but because of the offset between 3' and 5' positions, the spatial structure is that of a spiral ribbon.

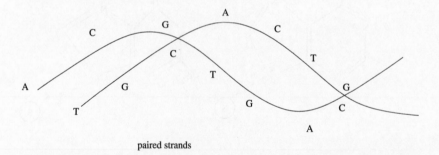

paired strands

Even the small portions of DNA – the genes – that code for proteins are not present in compact regions but, especially in the compact-nucleus eukaryotes, are interrupted by noncoding (and often highly repetitious) *introns*. The coding fragments – or *exons* – are also flanked by instructional subsequences, so that a small gene might look like: (5') upstream enhancer, promoter, start site, exon, intron, exon, poly-A site, stop site, downstream enhancer (3'). However, the vast remaining "junk DNA" – also riddled by fairly complex repeats (ALU, 300 base pairs; L1, very long; microsatellites, very short) – aside from its obvious mechanical properties, leading, e.g., to a supercoiled structure grafted onto the double helix, is of unknown function, and may be only an evolutionary relic.

The major steps in the DNA → protein sequence are well studied. Separation of the chains allows the exon–intron gene region of one of the chains to be read or *transcribed* to a pre-RNA chain of nucleotides (similar to the duplication of DNA needed in cell division) that differs from DNA by the substitution of *U* (uracil) for the *T* of DNA and by ribose (with a 2′-OH) for deoxyribose. The introns are then spliced out (by a signal still incompletely understood) to create messenger RNA, or *m*-RNA, which almost always (RNA can also be an end product) is itself read by transfer RNA, or *t*-RNA, which *translates*

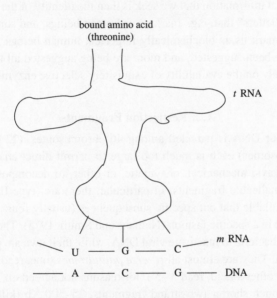

by setting up a specific amino acid for each base triplet of the *m*-RNA, or *codon* of the DNA, the amino acids then joining to form protein. The triplets code for 20 amino acids (as well as the start codon AUG at its first occurrence and stop codons UAA, UAG, UGA) when present in exons, and they come in four main varieties: nonpolar (hydrophobic), polar uncharged, + charged

$$HN - \underset{H}{\overset{R}{C}} - \overset{OH}{\underset{\underset{O}{\parallel}}{C}} \longrightarrow NH_2 - \underset{}{\overset{R_1}{CH}} - \underset{O}{\overset{}{C}} - NH - \overset{R_2}{CH} - \underset{O}{\overset{}{C}} - NH - \overset{R_3}{CH} - \overset{}{C} -$$

(basic), and – charged (acidic). Of course, there are always exceptions, and stop codons seem to be responsible as well for incorporation of crucial trace metals (selenium, zinc, etc.) into protein. Because there are 64 possible codons, there is a good deal of ambiguity, and the third member of the triplet is irrelevant in most cases. As we go along a DNA double strand (5×10^6 base pairs in *E. coli*, 3×10^9 – in 46 chromosomes – for us) there are six possible "reading frames" for triplets (3 times $5' \rightarrow 3'$ for either strand), and the correct one is selected by a start signal. The three-dimensional spatial or folding structure is important for the DNA and crucial for the resulting protein, but this is determined (precisely how is only partially clear – chaperonins, large protein templates, certainly help) by the one-dimensional sequence or primary structure, which is what we focus on.

The initial information that we seek is then the identity of the sequence of $\approx 3 \times 10^9$ "letters" that, e.g., mark us as human beings, and some of whose deviations mark us as biochemically imperfect human beings. Many techniques have been suggested, and more are being suggested all the time, but almost all rely on the availability of exquisitely selective enzymes.

1.2. Restriction Fragments

Although our DNA is parceled among 46 chromosomes, (22 pairs plus 2 sex chromosomes) each is much too large to permit direct analysis. There are many ways, mechanical, enzymatic, or other, to decompose the DNA into more malleable fragments. In particular, there are (type II) *restriction enzymes* available that cut specific subsequences (usually four, six, or eight letters long) in a specific fashion (Nathans and Smith, 1975). These enzymes are used by bacteria to inactivate viral DNA, while their own are protected by methylation. They are almost all *reverse palindromes* (one, read $5'-3'$, is the same as the other strand, read $3'-5'$), for reasons not agreed on. In this way, we create much shorter two-strand fragments, 25–500 Kb (kilobase pairs) depending, to analyze (the loose ends can also bind other loose ends created by the same enzyme to form recombinant DNA). In practice, many copies of the DNA are made, and only a portion of the possible cuts is performed, so that a highly diverse set of overlapping fragments is produced (see Section 1.3).

$$5' \dots GTT \uparrow AAC \dots 3' \qquad G \uparrow \underline{AATT} \quad C \qquad A \uparrow \underline{AGCT} \quad T$$
$$\qquad\qquad \uparrow \qquad\qquad\qquad\qquad \uparrow \qquad\qquad\qquad\qquad \uparrow$$
$$3' \dots CAA \uparrow TTG \dots 5' \qquad C \quad TTAA \uparrow G \qquad T \quad TCGA \uparrow A$$

Hpa 1   Eco R 1   Hind III

The fragments, which can be replicated or cloned in various ways, can then serve as a low-resolution signature of the DNA chain, or a large segment thereof, provided that they are characterized in some fashion. Of several in current use, the oldest characterization is the restriction-enzyme *fingerprint*: the set of lengths of subfragments formed, e.g., by further enzymatic digestion. These are standardly found, with some error, by migration in gel electrophoresis. Typically (Schaffer, 1983) we use the empirical relation $(m - m_0)(l - l_0) = c$, where m is migration distance and l is the fragment length, with m_0, l_0, and c obtained by least-squares fitting with a set of accompanying standard fragments (l_i, m_i): Define $c(m, l) = (m - m_0)(l - l_0)$ and minimize $Q = \sum_i [c(m_i, l_i) - c_{av}]^2$ to get m_0, l_0, and c estimates, and then compute by $l = l_0 + c_{av}/(m - m_0)$. What size fragments do we expect so that we can design suitable experiments? This is not as trivial as it sounds and will give us some idea of the thought processes we may be called on to supply (Waterman, 1983). A heuristic approach (Lander, 1989) will suffice for now.

It is sufficient to concentrate on one strand, as the other supplies no further information. Suppose the one-enzyme cut signal is a six-letter "word," ($5'$) $b_1 b_2 b_3 b_4 b_5 b_6$ ($3'$), and, as a zeroth-order approximation to the statistics of DNA, imagine that the letters occur independently and with equal probability, $p(A) = p(C) = p(T) = p(G) = 1/4$, at each site. Then, for each site, the probability of starting and completing the word to the right is simply $\frac{1}{4} \times \frac{1}{4} \times \frac{1}{4} \times \frac{1}{4} \times \frac{1}{4} \times \frac{1}{4}$,

$$p(b_1 b_2 b_3 b_4 b_5 b_6) = 1/4^6.$$

Suppose we have found one word and continue down the strand looking for the next occurrence. Assuming that $b_1 b_2 b_3 b_4 b_5 b_6$ cannot initiate a displaced version of itself, e.g., $b_5 b_6 \neq b_1 b_2$, we start after the word ends. Then the probability of not seeing a new word start for $l - 1$ moves but seeing one at the lth move is clearly the *geometric* distribution

$$p(l) = (1 - 1/4^6)^{l-1} \, 1/4^6$$

{or, because $1/4^6$ is very small, $p(l) \sim [(1/4^6)e^{-l/4^6}]$, the continuous *exponential* distribution}. The mean distance to the next word is then the mathematical expectation

$$\mu = E(l) = \sum_{l=0}^{\infty} \frac{1}{4^6} \left(1 - \frac{1}{4^6} \right)^{l-1} l.$$

On evaluation, $[\sum_{l=0}^{\infty} \alpha l (1 - \alpha)^{l-1} = -\alpha \frac{\partial}{\partial \alpha} \sum_{l=0}^{\infty} (1 - \alpha)^l = -\alpha \frac{\partial}{\partial \alpha} \frac{1}{\alpha} = \frac{1}{\alpha}]$, we have

$$\mu(b_1 b_2 b_3 b_4 b_5 b_6) = 4^6 = 4096.$$

The preceding argument will not hold for self-overlapping words, as the absence of a word starting at a given site slightly biases the possibilities for words starting at the next six sites, but because p is so small, this correlation effect is very small. We also have to distinguish between allowing two occurrences to overlap and not allowing it. In fact, a careful mathematical analysis (Guibas and Odlyzko, 1980) shows that the relation

$$\mu = 1/P$$

holds exactly for a long *renewal process*, one in which all the letters of a word are removed before we start counting again; here μ is the mean repeat distance from the beginning of the pattern and P is the probability that a renewal starts at a given site. Interestingly, this is precisely the situation that is said to exist with restriction enzymes – for a recognition site such as TAG CTA with self-overlap after moving four bases, a subsequence TAGCTAGCTA would be cut only once, whatever the direction of travel of the enzyme – there would not be enough left to cut a second time (the main reason seems to be that an enzyme needs something to hold onto and cannot work directly on a cut end). If this is the case, the mean repeat distance will change. In this example, we still have the basic $p(\text{TAGCTA}) = 1/4^6$, but the unrestricted p at site n is composed of either a repeat, say at site n, or a repeat at site $n - 4$, followed by the occurrence of GCTA to complete the TA pair: $p = P + 4^{-4}P$. Hence $\mu = 1/P = (1 + 4^{-4})/p = 4^6 + 4^2 = 4112$. More generally, we find

$$\mu = 4^6(1 + e_1/4 + \cdots + e_5/4^5),$$

where $e_i = 1$ for an overlap at a shift by i sites, otherwise $e_i = 0$.

The relevance of the above discussion in practice is certainly marginal, as the significance of such deviations is restricted to very short fragments, which are generally not detected anyway. However, the assumption of independent equal probabilities of bases is another story. To start with, these probabilities depend on the organism and the part of the genome in question, so that we should really write instead

$$p(b_1 \cdots b_6) = p(b_1) \cdots p(b_6),$$

and this can make a considerable difference, which is observed. To continue, we need not have the independence $p(bb') = p(b)\, p(b')$; rather,

$$g(bb') = p(b\, b')\, /\, p(b)\, p(b')$$

measures the correlation of successive bases – it is as low as $g(CG) \sim 0.4$. If this successive pair correlation or Markov chain effect is the only correlation present, we would then have

$$p(b_1 \cdots b_6) = p(b_1) \cdots p(b_6)\, g(b_1b_2)\, g(b_2b_3)\, g(b_3\, b_4)\, g(b_4\, b_5)\, g(b_5\, b_6),$$

and this effect too is observed, although some frequencies are more strongly reduced, implying correlations at intersite separations as large as ten. We will examine this topic in much greater detail in Section 3.

1.3. Clone Libraries

As mentioned, we typically start the analysis of a genome, or a portion thereof, by creating a library of more easily analyzed fragments that we hope can be spliced together to recover the full genome. These fragments can be replicated arbitrarily, or cloned, by their insertion into a circular *plasmid* used as a blueprint by bacterial machinery, by other "vectors," and by DNA amplification techniques. Each distinct fragment is referred to as a clone, and there may be practical limits as to how many clones can be studied in any attempt to cover the full portion – which we simply refer to as the genome. Assume a genome length (in base pairs) of G, typical length L of a clone, and N distinct clones created by mechanical fragmentation of many copies, so they might start anyplace. How effectively can we expect to have covered the genome, i.e., are there still "oceans" between "islands" of overlapping clones? For a quick estimate, consider a base pair b at a particular location. The probability of its being contained in a given clone c is obtained by moving the clone start over the G positions, only L of which contain b:

$$P(b \in c) = L/G,$$

so that

$$P(b \notin c) = 1 - \frac{L}{G}.$$

Hence $P(b \notin any \text{ clone}) = (1 - \frac{L}{G})^N \sim e^{-LN/G}$, so that the expected fraction of the genome actually covered is the "coverage" (Clarke and Carbon, 1976):

$$f = 1 - e^{-c}, \qquad c = LN/G;$$

equally often, c itself is referred to as coverage. Note that if the clone starts are not arbitrary, but "quantized" by being restriction sites, this on the average just changes the units in which G and L are measured.

Let us go into detail; see, e.g., Chapter 5 of Waterman (1995). Suppose first that we are cutting a single molecule with a single restriction enzyme. Not all clones have exact length L, and if a clone is inserted into a plasmid or other vector for amplification, it will be accepted only within some range

$$l \leq L \leq U.$$

A clone of length L will be produced by two cuts of probability p (e.g., $\sim 1/4000$ for Eco R1), separated by L no-cuts, a probability of $(1 - p)^L \sim e^{-Lp}$. A located base pair b can occur at any of L sites in such a clone, a net probability for $b \in C$ of $p^2 L e^{-Lp}$. Hence, imagining continuous length L to convert sums to integrals, we find that the probability of that b is in some clonable fragment – i.e., the fraction of G covered by cloned fragments – is given by

$$
\begin{aligned}
f &= \int_l^U p^2 L e^{-pL}\, dL = -p^2 \frac{\partial}{\partial p} \int_l^U e^{-pl}\, dL \\
&= -p^2 \frac{\partial}{\partial p} \frac{1}{p} (e^{-pl} - e^{-pU}) \\
&= (1 + pl) e^{pl} - (1 + pU) e^{-pU},
\end{aligned}
$$

close to unity only for pl small, pU large, which is never the case in practice.

A clone library should do a better job of covering the genome, and we can accomplish this by using, e.g., a 4-cutter on many copies of the genome, but stopping at partial digestion. Suppose the digestion sites occur at mean frequency p – fixed in the genome – but only a fraction μ are cut, giving a large distribution of cut sites for a system of many double strands. For a quick estimate, again with an acceptance range of l to U, the expected number of restriction sites between two ends of a clonable fragment is between pl and pU. If μ is the fraction cut, the probability that such a fragment, starting at a given restriction site, actually occurs is at least $\mu^2 (1 - \mu)^{pU}$. However, there are $\sim Gp$ restriction sites all told, each the beginning of $p(U - l)$ fragments. The estimated number of molecules required for picking up all of these is therefore of the order of

$$
\# = Gp^2(U - l)/\mu^2(1 - \mu)^{pU},
$$

and many more will certainly do it. As an example, for *E. coli*, $G = 5 \times 10^6$, cutting with Eco $R1$, $p = 4^{-6}$, at $\mu = 1/5$, and cloning with pJC74, $l = 19 \times 10^3$, $U - l = 17 \times 10^3$ yields $\# \sim 1.8 \times 10^6$, which is much smaller than the normally available 2×10^9 molecules. For human DNA fragments, large cloning vectors are used to create large numbers of identical molecules. [The problem of splicing together fragments from the soup resulting from such cutting procedures can be avoided if the rapid shuffling can be avoided. For this purpose, the ideal would be to focus on a single molecule with an undisturbed sequence. A developing technique (Schwartz et al., 1993) does this by uniform fluorescence – staining DNA, stretching it out by fluid flow, and fixing it in a gel. Application of a restriction enzyme then puts gaps in the pattern of fluorescence as the fragments contract a bit, allowing optical measurement of the intensity and hence the length of the fragments. For algorithmic aspects, see, e.g., Karp and Shamir (2000). Reliability is increasing, but this has not yet led to the extensive sequencing available from the techniques to be described.]

Let us return to the characterization of a subfragment of DNA, say some fraction of a megabase. Using a single restriction enzyme, we get a set of fragments of various lengths, cut at both ends by the enzyme; these fragments can be analyzed at leisure, and at least named meanwhile. However, in what order do they occur? If known, this defines a *restriction map*; to produce it, we can use the method of double digestion, first by two enzymes, A and B, separately, and then together. In other words, we have unordered lengths $(a_1, \cdots, a_n)$ by A cutting, $(b_1, \cdots, b_n)$ by B cutting, and $(c_1, \cdots, c_t)$ by A and B cutting. What site locations are compatible with this data? Aside from its being a hard problem to find a solution (technically, NP hard), there may be many mathematical solutions that have to be distinguished from each other. A major contribution to this uncertainty comes from typical length-measurement uncertainties, say of Δ sites.

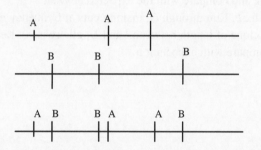

Suppose (Waterman, 1983; Lander, 1989; Selim and Alsultan, 1991) that p_A is the probability of an A cut at a given (multiple) site and p_B the probability of a B cut. The probability that a pair of cuts is indistinguishable – an effective AB coincidence – is the probability of an A cut at a given site, and then of a B cut at one of Δ sites, or $\Delta p_A p_B$. Segments terminated by AB coincidence on each side can be freely permuted without the sets of A, B and A or B fragment lengths being changed. The expected number of such ambiguous cuts in a chain of length L is

$$s \sim L\Delta p_A p_B.$$

There are then $s!$ orderings of fragments that are indistinguishable, which can be very large for large L. For example, for the full human genome,

$L \sim 3 \times 10^9$, and two 6-site restriction enzymes, we would have $s \sim 3 \times 10^9 \times 50/(4000)^2 \sim 10^4$. For a single chromosome, this is reduced to 200, but for as large as a megabase, it is reduced to only 0.3. For such lengths, decent algorithms, both deterministic and stochastic, exist. Another strategy is to make use of multiple, e.g., ≥ 3 enzymes, complete digest mapping to reduce ambiguity, and work along these lines has been reported (Fasulo et al., 1999).

Assignment 1

1. Consider a 2-letter alphabet, say 0 (purine) and 1 (pyrimidine). By inspecting the 32 numerically ordered 5-digit numbers, 00000 to 11111, placed in sequence, find the renewal frequency of 00 and 01. Are these consistent with your expectation?

2. Choose reasonable parameters G, L, and N. Then fix the base pair b in a G-site double strand and randomly choose the center of a length L clone on $G - L$ sites. Check 1 if $b \in$ clone, 0 if not, and do N times to find f, the fraction of the genome actually covered. Average over many repeats, and compare with the expected answer.

3. Again fix b. Run through G, making cuts at frequency p. Check 1 if b is in a clone of length between l and U, otherwise 0. Repeat, average, and compare with expectation.

2

Recomposing DNA

2.1. Fingerprint Assembly

We now leave the world of estimates and enter that of exact results, but for model situations. The chain to be analyzed is imagined to be present as a set of cloned subchains with substantial but unknown overlap. In this section, we characterize a member of a clone by an ordered set of restriction fragments, or just a set of restriction-fragment lengths, or a set of lengths of special restriction fragments (e.g., those with certain characteristic repeats) called the *fingerprint* of the clone. We have, in many cases, a library of randomly chosen pieces of DNA or a section of DNA, each with a known fingerprint. Can we order these to produce a *physical map* of the full sequence? To examine the degree to which this is feasible (Lander and Waterman, 1988), let us first expand our notation. G will denote genome length (in base pairs, bp), L the clone length, N the number of distinct clones available, $p = N/G$ the probability of a clone's starting at a given site, and $c = LN/G = Lp$ the redundancy, the number of times the genome is covered by the aggregate length of all clones. In addition, and crucially, we let T be the number of base pairs two clones must have in common to declare reliably that they overlap, $\theta = T/L$ the overlap threshold ratio, and $\sigma = 1 - \theta = (L - T)/L$; multiple overlap is fine. Note again that if the clones are produced by mechanical shearing, they can indeed start anyplace, but if produced by enzymatic digestion, "location" is in units of mean interrestriction site distance; this will not affect our results, in which only length ratios appear. We will follow the elementary treatment of (Zhang and Marr, 1992).

Clones that are declared as overlapping build up a sequence of contiguous clones, *an island*, or more accurately, an *apparent* island, as not all overlaps are detected. An island of two or more clones is called a *contig*, and the gaps between islands are termined *oceans*. To see how effectively we can pick up the full chain in question, we need the statistics of the islands.

1. First, how many islands on the average, and how big? We use the above notation, with the tacit assumption, to be removed in Section 2.5, that all clones

13

have the same length and the same overlap threshold. Now we start at one end of the genome and move through it. The probability of starting a clone at

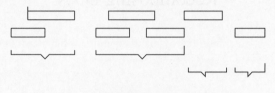

apparent islands

a given base pair is p; the probability of starting one but *not* detecting a next overlap is $p(1 - p)^{L-T} = p(1 - p)^{c\sigma/p} \sim pe^{-c\sigma}$ (no clone can start at the next $L - T$ base). Because the number of (apparent) islands is just the number of times we leave a clone without picking up another clone by overlap, the

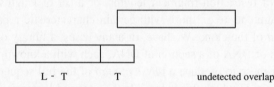

L - T   T   undetected overlap

expected number is $E(N_I) = Gpe^{-c\sigma}$, or

$$E(N_I) = Ne^{-c\sigma}.$$

(There is a slightly subtle point in this argument: It is assumed that the event that an island that starts its last clone at x is independent of one starting its last clone at y. This holds because there are many identical copies of the genome or segment under study, so that the two clones to be compared come from independent segments; in other contexts, the independence is only an approximation.)

2. According to the above argument, the probability that a given clone, one starting at a specified site, terminates an island is $e^{-c\sigma}$, that it does not, $1 - e^{-c\sigma}$. Hence the probability that a given island has exactly j clones is

$$P_{I,j} = (1 - e^{-c\sigma})^{j-1} e^{-c\sigma}$$

(indeed $\sum_{j=1}^{\infty} P_{I,j} = 1$, as it should). Multiplying $P_{I,j}$ by the expected number of islands, $E(N_I)$, gives us the expected number of j-clone islands, i.e.,

$$E(N_{I,j}) = Ne^{-2c\sigma}(1 - e^{-c\sigma})^{j-1},$$

and the mean number of contigs, islands that have more than just one clone, is

$$E(N_{con}) = Ne^{-c\sigma} - Ne^{-2c\sigma}.$$

Also, of course, the mean number of clones per island is $E(j) = \sum_j jP_{I,j}$, or

$$E(j) = e^{c\sigma}.$$

Note that the expected number of islands in units of the maximum number of islands, $G/L = N/c$, becomes

$$\frac{L}{G}E(N_I) = ce^{-c\sigma},$$

a fairly sensitive (universal) function of the required overlap θ. Sample

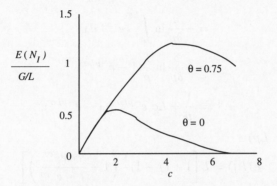

values of G/L for *E. coli* and human, phage-derived, and yeast-derived clones (termed YAC for yeast artificial chromosone), are given (kb stands for kilo-base pair, Mb for megabase pair):

	Phage (15 kb)	Yeast (1 Mb)
E. coli	270	4
Human	200,000	3000

3. A better idea of the rate of progress lies in the actual island length as measured in base pairs. An island of j clones will have a base-pair length of

$$L_j = x_1 + x_2 + x_3 + \cdots + x_{j-1} + L,$$

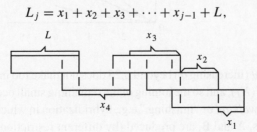

where, proceeding from right to left, x_i is the portion of clone i that does not overlap the next clones and, as we have seen, occurs with a probability of

$$P(x_i) \propto (1 - p)^{x_i} \sim e^{-px_i}$$
$$\text{for } 0 < x_i \le L - T = L\sigma.$$

Now, at fixed j, because all x_i have the same distribution,

$$E(L_j) = L + (j - 1) E(x).$$

If x is represented as continuous,

$$\begin{aligned}
E(x) &= \int_0^{L\sigma} x\, e^{-px}\, dx \Big/ \int_0^{L\sigma} e^{-px}\, dx \\
&= -\frac{\partial}{\partial p} \ln \int_0^{L\sigma} e^{-px}\, dx \\
&= -\frac{\delta}{\delta p} \ln \frac{1}{p}(1 - e^{-pL\sigma}) \\
&= \frac{1}{p} + L\sigma\, e^{-pL\sigma}/(1 - e^{-pL\sigma}),
\end{aligned}$$

so that $(c = Lp)$

$$E(L_j) = L\left[1 + (j - 1)\frac{1}{c}\left(1 - \frac{c\sigma e^{-c\sigma}}{1 - e^{-c\sigma}}\right)\right].$$

Averaging over j, using $E(j) = e^{c\sigma}$, we obtain the mean island length in units of clone length:

$$\frac{1}{L} E(L_I) = 1 - \sigma + \frac{1}{c}(e^{c\sigma} - 1).$$

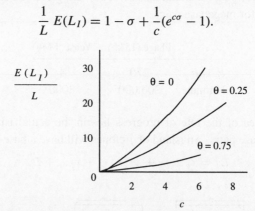

Lowering θ (increasing σ) beyond 0.25 does not make too much difference in $E(L_I)$ or $E(N_I)$, and so the joining of the remaining small oceans is done by *chromosome walking* or "finishing," e.g., hybridization in which two different sets of islands, A and B, are produced (by different restriction enzyme sets). A member of A is used to bind to one of B, then to another of A, etc., until a complete overlapping set is produced. More systematic walking procedures extend an island by ~500 bp each step until one meets the next island. We do this by using the island as a primer in a polymerase chain reaction (PCR) (Saiki, 1988) to build the following portion of the genome, the near end of which is then laboriously sequenced by the Sanger technique (Sanger et al., 1977). In fact, in high-resolution studies, a section is not regarded as finished until each site is identified from three clones. A balance must be sought between the cost of high coverage with few unfinished sections and the cost of finishing (Czabarlia et al., 2000).

In declaring overlap, a practical consideration concerns the rate of false positives: overlap attributed to nonoverlapping clones, because they have a common signature, to within experimental error. In particular (Lander and Waterman, 1988), suppose restriction-subfragment lengths x and $x(1 - \beta) \leq y \leq x(1 + \beta)$ are taken as matching. If each is chosen randomly from an exponential distribution $(x \geq 0)$ $Pr(x) = \lambda e^{-\lambda x}$ (a reasonable assumption), the chance that such subfragments will be seen as matching is $\int_0^\infty \int_{x(1-\beta)}^{x(1+\beta)}$ $(\lambda e^{-\lambda y} \, dy) \lambda e^{-\lambda x} \, dx = 2\beta/(4 - \beta^2) \sim \beta/2$. Now, as an example, suppose the common signature is a set of k subfragments of a full restriction map. Random overlap then has a probability of $4(\beta/2)^k$ (two maps have four orientations), and $\sum_k^\infty 4(\beta/2)^l = 4(\beta/2)^k(1 + \beta/2)$ for at least k matches. This determines the statistics to be used in conjunction with the overlap parameter $\theta \sim k/n$ for a mean n fragments per clone.

Let us return briefly to drop the assumption of fixed L and T. The dominant effect is the distribution of $1 - T/L = \sigma$, say $\rho(\sigma)$, suggesting that the basic probability $e^{-c\sigma}$ that a given clone terminates an island be replaced with

$$w(c) = \int \rho(\sigma) e^{-c\sigma} \, d\sigma \equiv \langle e^{-c\sigma} \rangle.$$

Quite generally, the average $\langle e^{-c\sigma} \rangle \geq e^{-c\langle\sigma\rangle}$ (the exponential is convex) but in more detail,

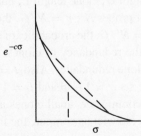

we have the *cumulant expansion*

$$\langle e^{-c\sigma} \rangle = \exp\left[-c\langle\sigma\rangle + \frac{c^2}{2}\text{var}(\sigma) + \cdots \right],$$

where $\mathrm{var}(\sigma) = \langle(\sigma - \langle\sigma\rangle)^2\rangle$, which is a first estimate for corrections for a tight distribution $\rho(\sigma)$. The situation is of course not that simple, as we shall see in Section 2.5. An important generalization in another direction is to the biologically relevant case of inhomogeneous density $c(t)$ of clones starting at t. This has been carried out for the subcase $\theta = 0$ (Karlin and Macken, 1991), and more generally we conclude (Port et al., 1995) that, for example, the expected number of apparent islands is

$$E(N_I) = \int_0^{G/L-\sigma} Lc(t)\, e^{-\int_t^{t+\sigma} c(s)\,ds}\, dt,$$

where location is measured in clone-length units.

2.2. Anchoring

A second "traditional" method of sequentially ordering clones to build up large islands, eventually to be converted to a full physical map of the genome, is known as *anchoring* (Arratia et al., 1991; Torney, 1991). Typical anchors are relatively short subsequences – *sequence-tagged sites* (STSs) – occurring at very low frequency (perhaps once) in the genome, which are used as *probes* to bind a complementary subsequence somewhere in the middle of a clone and so characterize it without the messy restriction-fragment decomposition, but at the cost of building up a library of anchors. On the other hand, STS signatures allow for ready pooling of data from different research groups. Two clones are recognized as contiguous if they bind the same probe, and for many clones at the same probe, only the two extending furthest to left and right need be retained. We analyze this sequencing system in a fashion similar to that for fingerprinting, but of course with its own special characteristics. Again, we follow Zhang and Marr (1994a).

The notation is an extension of that used in Section 2.1: we have the fixed parameters of genome length G; clone length L; number of clones N; probe length $M < L$; number of probes N'; $p = N/G$, the probability of a clone's starting at a given site; $\alpha = N'/G$, the probability of a probe starting at a given site; $c = NL/G = pL$, the redundancy of clones; and $d = N'L/G = \alpha L$, the maximum anchored clone redundancy. Also, $t = M/L$ is the relative size of probe to clone, and we set $q = 1 - p$ and $\beta = 1 - \alpha$. We again maintain discreteness at first to accommodate small clones and probes but mainly to simplify the presentation. Now let us compute the following.

1. The expected number of islands. We focus on the left end of an island; a clone starts at, say, 0, its *first* completely contained probe at one of $i = 0, \ldots, L - M$, and there is no clone to the left of 0 and anchored by

the same probe, i.e., none starting at $i + M - L, \ldots, -2, -1$. Hence the probability that an island starts at 0 is

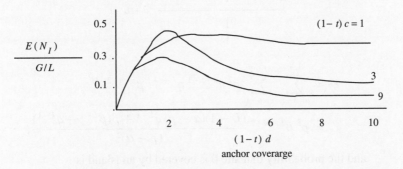

$$P(I) = \sum_{i=0}^{L-M} (pq^{L-M-i})(\alpha\beta^i)$$

$$= p\alpha \frac{\beta^{L-M+1} - q^{L-M+1}}{\beta - q}$$

(depending on $L - M$ alone) and then the desired

$$E(N_I) = GP(I).$$

If we drop the discreteness by letting $p, \alpha \to 0$ at fixed *coverages* $c = pL, d = \alpha L$, then $(1 - p)^x \to \exp - px = \exp - c(x/L), (1 - \alpha)^y \to \exp - d(y/L)$, so that

$$E(N_I) = N\alpha \frac{e^{-d(1-t)} - e^{-c(1-t)}}{c - d},$$

and we would clearly want to reduce the relative probe size t (subject to controlling false positives) in order to increase the effective c and d. High c and d of course reduce the expected number of islands.

2. The fraction of genome covered by islands. Define ρ_i as the probability that a given position is covered by i islands (i can exceed 1 because an overlapping pair of clones need not have a common anchor, but we can see that $i \leq 4$ for $L > 2M$). Then we compute ρ_0, the probability that a given site is not covered by an island, in terms of that of ξ_0, the probability that no clone covers the site, ξ_1 that it is covered by one

unanchored clone, and ξ_2 that it is covered by two or more unanchored clones:

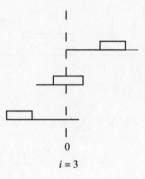

$i = 3$

a. no clone; $\xi_0 = q^L$;
b. one unanchored clone,

$$\xi_1 = L(pq^{L-1})\beta^{L-M+1};$$

c. more than one unanchored clone, the left end of the leftmost unanchored clone covering 0 at u, $1 - L \leq u \leq -1$; the right end of the rightmost unanchored clone covering 0 at v, $u + L \leq v \leq L - 1$. Hence

$$\xi_2 = \sum_{u=1-L}^{-1} \sum_{v=u+L}^{L-1} p^2 \, q^{u+L-1} \, q^{L-v-1} \, \beta^{v-u+2-M}$$

$$= p^2 \beta^{L-M+2} \frac{(L-1)(\alpha-p)\,q^{L-1} + \beta(\beta^{L-1} - q^{L-1})}{(p-\alpha)^2},$$

and the probability that site 0 is covered by an island is

$$1 - \rho_0 = 1 - q^L - Lpq^{L-1}\beta^{L+M+1} - \xi_2,$$

reducing in the continuous limit to

$$1 - \rho_0 = 1 - e^{-c} - ce^{-[c+(1-t)d]}$$

$$+ \frac{c^2(c-d+1)}{(c-d)^2} \, e^{-[c+(1-t)d]} - \frac{c^2}{(c-d)^2} \, e^{-(1-t)d}.$$

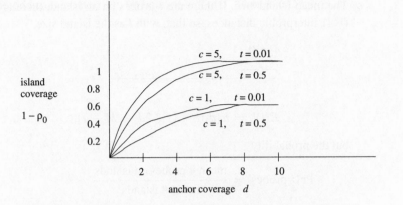

anchor coverage d

Note that, for large d, $\rho_0 = e^{-c}$, a typical Poisson process result.

3. The expected size of an island. We compute this in stages:

a. The mean distance between adjacent probes on the same island and hence on the same clone. We fix probe P with its right end at 0, and then take the right end of the rightmost *clone* anchored by P

as y, with probability $\propto q^{L-M-y}$. Given these, the probability that the right nearest-neighbor probe ends at x is $\propto \beta^{x-M}$ (there is no previous probe). Hence the mean distance between probes on the *same* island is

$$d_1 = \left(\sum_{0 \le x \le y \le L-M} x \beta^x q^{-y} \right) \bigg/ \left(\sum_{0 \le x \le y \le L-M} \beta^x q^{-y} \right)$$

$$= \frac{1}{\frac{q}{\beta} - 1} - \frac{L - M + 1}{\left(\frac{q}{\beta}\right)^{L-M+1} - 1}.$$

b. The mean distance from the right most probe to the right end of an island, and hence to the right end of the last clone is seen to be

$$d_2 = \left(\sum_{0 \le y \le L-M} y q^{-y} \right) \bigg/ \left(\sum_{0 \le y \le L-M} q^{-y} \right) = \frac{L-M+1}{1-q^{L-M+1}} - \frac{1}{1-q}.$$

c. The mean island size. If there are i probes on an island, there are $i - 1$ interprobic distances, so that, with l as the island size,

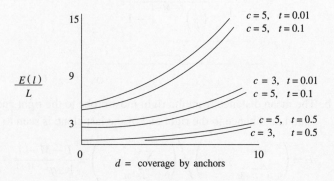

$$E(l)|_i = E[d_1(i - 1) + 2d_2 + M - 1],$$

but the probability

$$\Pr(i \text{ probes}) = \frac{\text{mean \# probes on islands}}{\text{mean \# islands}}$$

$$= \frac{N'[1 - \text{prob (no probes on a clone)}]}{E(N_I)}$$

$$= N'(1 - q^{L-M+1}) / E(N_I),$$

giving the net

$$E(l) = d_1 \left[\frac{N'(1 - q^{L-M+1})}{E(N_I)} - 1 \right] + 2d_2 + M - 1.$$

Extension to inhomogeneous clone density has been carried out for anchored islands as well (Schbath et al., 2000). For example, if $f_x(L)$ is the length distribution of clones ending at x, $F_x(L) = \int_L^{\infty} f_x(L') \, dL'$, and $J(x, L) = \exp -c \int_L^{\infty} F_{x+y}(y) \, dy$, then the mean number of anchored islands is found to be

$$E(N_I) = \alpha p \int_0^G \int_0^{\infty} F_x(L') \, J(x - L', L') \, e^{-\alpha L'} \, dL' \, dx,$$

which reduces to our above result when $f_x(L') = \delta(L' - L)$. One situation in which clone length is not tightly distributed is that of radiation hybrid mapping (Goss and Harris, 1975; Cox et al., 1990; Slomin, 1997), in which rodent–human cells are created from radiation-broken human DNA fragments to serve as sources of clones.

An extreme version of the anchoring technique, "hybridization mapping" (Poustka, 1986), has come under increased scrutiny (Kim and Segre, 1999; Mayraz and Shamir, 1999). It is that in which the coverage by anchors alone is very high, producing an overlapping population of nonunique location specifiers, and hence not representable by the above analysis. Because the anchors themselves cover the genome, their statistics make it possible to recognize repeated subsequences and save their incorporation until the end of the process. Also the problem of noise during fragment construction and recognition is reduced by the numerous anchors that connect two fragments, as well as the concomitant interprobe distance information.

2.3. Restriction-Fragment-Length Polymorphism (RFLP) Analysis

Suppose we have obtained a physical map of the genome, or a chromosome, . . . , of an individual by means of a complete set of overlapping clones together with their restriction-fragment-length fingerprints, now a traditional if very lengthy activity. The net effect is then a sequence of fingerprint markers or signposts. When DNA appears altered, e.g., by a mutation or simply by transmission of a previous alteration, this alteration can then be inherited. However, let us recall that the chromosomal DNA contributed by one human parent is not simply one of the pair of homologous chromosomes in each cell. Rather, during the process of meiosis resulting in the (halved) chromosomal complement of a gamete, homologous strands swap segments. For any transmitted alteration to be viable, the whole length of the gene – coding and regulatory regions – must hang together during the crossing over, a minimum amount, say x, of altered DNA. To keep track of such an alteration, we want to have some marker within this minimum distance, so that the marker will transfer with the gene. Distances on DNA are often measured operationally in *genetic linkage* units: a centimorgan (cM) or a "map unit" is the separation distance of two loci that are separated 0.01 of the time during the chromosome crossover period (the human genome has $\sim$3300 cM) and a reasonable requirement might be that every major gene locus is within 10 cM of a marker (clearly, 1 cM $\sim$3 $\times$ $10^9/3300 \sim 10^6$ bp; the transformation from physical to genetic distance is not really uniform, but we will neglect this fact, to leading approximation).

An altered gene – an *allele* of the reference DNA – then carries with it an altered set of restriction-fragment lengths, termed a *restriction-fragment-length polymorphism* (RFLP). To satisfy the above criterion, suppose first that n markers are placed on a genome, or chromosome, of map length L cM. What is the probability $P_{n,x}(L)$ that the whole sequence is covered by the n intervals of length x cM, centered at the markers? (Ideal placement would allow $n = L/x$ to do the trick perfectly (Lange and Boehnke, 1982; Bishop et al., 1983).)

Except for end effects that are easily taken care of, this is the same as the probability that n ordered points $0 \le y_1 \le y_2 \cdots \le y_n \le L$ produce intervals of only $\le x$. The volume swept out by the points $(y_1, \ldots, y_n)$ of the n-dimensional space without restriction is of course $L^n/n!$; the restricted volume is now

$$V_{n,x}(L) = \int \cdots \int_{\substack{0 \le y_1, \cdots 0 \le y_{i+1} - y_i \le x \\ 0 \le L - y_n \le x}} dy_1, \ldots, dy_n,$$

$$P_{n,x}(L) = (n!/L^n)V_{n,x}(L).$$

A device we often use is that of a *generating function*. We define

$$Q_{n,x}(t) = \int_0^\infty e^{-tL} V_{n,x}(L)\, dL$$

(the Laplace transform, the standard moment-generating function of probability). Switching to variables $z_1 = y_1, z_2 = y_2 - y_1, \ldots, z_n = y_n - y_{n-1}$, $z_{n+1} = L - y_n$, this can be written as

$$Q_{n,x}(t) = \int_0^x \cdots \int_0^x e^{-t(z_1 + \cdots + z_{n+1})}\, dz_1 \cdots dz_{n+1}$$

$$= \left(\int_0^x e^{-tz}\, dz \right)^{n+1} = \left(\frac{1 - e^{-tx}}{t} \right)^{n+1}.$$

This is very simple, but how do we return to $P_{n,x}(L)$? One way involves inverting the Laplace transform; on consulting tables, we find that,

$$P_{n,x}(L) = \sum_{0 \le j < L/x} (-1)^j \binom{n+1}{j} \left(1 - \frac{jx}{L} \right)^n,$$

which is numerically but not analytically useful without further transformations. A better way is by direct asymptotic evaluation of the inverse Laplace transform. The formula for the inverse Laplace transform is most readily obtained from that of the Fourier transform, which extracts frequency components and then puts them together again:

$$\text{if } h(x) = \frac{1}{2\pi} \int_{-\infty}^{\infty} e^{-ixy} g(y)\, dy$$

and $g \to 0$ rapidly as $|y| \to \infty$,

$$\text{then } g(z) = \int_{-\infty}^{\infty} e^{ixz} h(x)\, dx.$$

Now if we let $g(y) = e^{-Ky} f(y)\theta(y)$, where

$$\theta(y) = \begin{cases} 0 & y < 0 \\ 1 & y \geq 0 \end{cases}$$

is the unit step function, then $e^{-Ky} f(y)\theta(y) = \frac{1}{2\pi} \int_{-\infty}^{\infty} \int_{-\infty}^{\infty} e^{-Kz} f(z)\theta(z)$ $e^{-ikz}\, dz\, e^{iky}\, dk$, or if $t = ik + K$ and $y \geq 0$,

$$f(y) = \frac{1}{2\pi i} \int_{K-i\infty}^{K+i\infty} e^{ty} \left[\int_0^\infty f(z) e^{-tz}\, dz \right] dt,$$

the Laplace transform inversion formula.

Here then

$$V_{n,x}(L) = \frac{1}{2\pi i} \int_{K-i\infty}^{K+i\infty} e^{tL} \left(\frac{1 - e^{-tx}}{t} \right)^{n+1} dt,$$

or, in terms of the coverage parameter defined for present purposes as $c = nx/L$,

$$V_{n,x}(L) = \frac{1}{2\pi} \int \left[e^{\frac{tx}{c}} \left(\frac{1 - e^{-tx}}{t} \right) \right]^n \frac{1 - e^{-tx}}{t}\, dt.$$

In general,

$$\lim_{n \to \infty} \left| \int a(t)^n b(t)\, dt \right|^{1/n} = \max_t |a(t)|.$$

In the present case, $a(t)$ is stationary at a real value of t, which is minimum in the real direction, but maximum in the imaginary direction. It is given by

$$\frac{x}{c} - \frac{1}{t} + \frac{x e^{-tx}}{1 - e^{-tx}} = 0,$$

so that, if tx is large, then $(tx/c) - 1 = -[tx/(e^{tx} - 1)] \to 0$ or $t = c/x$, and [by means of Stirling's approximation in the form $(n!)^{1/n} \sim n/e$]

$$P_{n,x}(L)^{1/n} \sim \frac{ne^{-1}}{L} V_{n,x}(L)^{1/n}$$

$$\sim \frac{c}{x} e^{-1} e \frac{1 - e^c}{c/x} = 1 - e^{-c}.$$

Hence

$$P_{n,x}(L) \sim \exp - (ne^{-c}).$$

Thus to have a reasonable probability of full coverage, we need $n \sim e^c$. In particular, for the full human genome, with $x = 20$ cM for 10-cM resolution, $L/x \sim 3300/20 \sim 165$, so that $\ln c = c/165$, leading to the requirement $c \sim 7$ or $n \sim 1170$. Substantially higher values of c will get the probability very close to 1.

In further detail, we may study the mean and the standard deviation of the proportion C_n of the genome covered by n markers randomly placed on r chromosomes of lengths $l_1, l_2, \ldots, l_r > x$ cM strung together, with $\sum l_i = L$. It can be shown (Robbins, 1944), and is certainly reasonable, that the expectation of the uncovered part is given by

$$E(1 - C_n) = p_{1,n},$$

the probability that a randomly placed location is not covered by n randomly placed intervals, uniform over the whole genome. Similarly,

$$E[(1 - C_n)^2] = p_{2,n},$$

the probability that two randomly placed sites are not covered by n randomly placed intervals. Let us look at $p_{1,n}$. There are two possibilities: (1) a random site falls inside a chromosome, at least $x/2$ cM from each end; this has probability (inside length/length) $= (L - rx)/L = 1 - rx/L$. The probability that this site is not covered by any of the n intervals is of course $(1 - x/L)^n$; or (2) a site falls $y < x/2$ cM from a chromosome end with probability rx/L and is not covered by n intervals with probability $[1 - (y + x/2)/L]^n$. Summing y from 0 to x/L and adding the three contributions, we obtain

$$1 - E(C_n) = p_{1,n} = \left(1 - \frac{rx}{L}\right)\left(1 - \frac{x}{L}\right)^n$$

$$+ \frac{2r}{n+1}\left[\left(1 - \frac{x}{2L}\right)^{n+1} - \left(1 - \frac{n}{L}\right)^{n+1}\right],$$

or, because $(1 - x/L)^n = e^{-c}$ in the continuous limit,

$$E(C_n) = 1 - \left(1 - \frac{rx}{L}\right)e^{-c} - \frac{2r}{n+1}(e^{-c/2} - e^{-c}),$$

and for $n \sim e^c$, this is again dominated by the e^{-c} term.

With the coverage under control, we can take advantage of RFLPs in both medical and forensic directions. For the moment, we just note that RFLPs can be generated both by variation in the distribution of restriction sites, by a very important subcase, the variation in number of *tandem repeats* (VNTR), and more recently by repeats of the very short microsatellites (with no genetic implications). The tandem repeats are genes that were initially repeated by recombination of pieces and then accumulated a large distribution by crossing over; their multiple nature allows them to be preferentially selected by hybridization. A typical situation involving the transmission of an altered gene (here, one C is replaced by a T) is shown (Botstein et al., 1980).

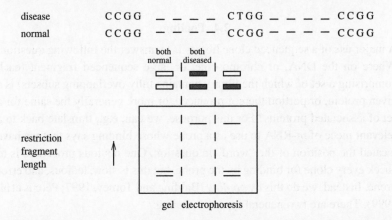

A CCGG cutter acting on the two-strand sections of sister chromosomes shown (white for normal, gray for diseased) produces three characteristic restriction-fragment signatures. If the disease gene is recessive, their (two copies on the two strands are necessary) progeny, their disease state, and their restriction fingerprints might go as

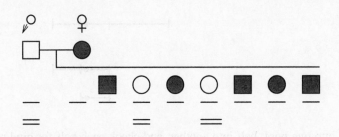

and so the genetic risk is readily assessed.

Assignment 2

1. In fingerprint assembly, suppose that the clone density $c(t)$ varies linearly from 0 at each end of the genome to a maximum at the center.

Compare $E(N_I)$ with that obtained by using the average $c(t)$ and explain the result.

2. In the anchoring technique, find the probability that no gap between islands exceeds b base pairs.

3. The problem of asymptotic RFLP coverage ended up as a saddle-point estimation. Find the coefficient in front of the exponential.

2.4. Pooling

A major use of a sequenced clone library is to answer the following question: Where on the DNA, or chromosome, or large sequenced fragment (each comprising a set of which the library contains fully overlapping subsets) is a given protein, or portion thereof produced, or more generally the same for a set of associated proteins? For this purpose, we can, e.g., translate back to a relevant piece of m-RNA to use as a probe whose binding says that we have located the position of the "word" in question. One obvious procedure is to check every clone for binding by the probe, but this is slow, tedious, and error prone. Instead, we do this by *pooling* (Balding and Torney, 1997; Percus et al. 1999). There are two general types of pooling:

1. Adaptive pooling, prototypically patterned often on the traditional technique of detecting a (heavier) counterfeit coin. We mix half the clones

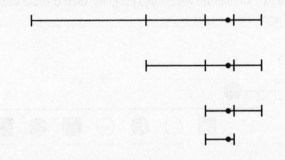

into one pool, half into another, and check each half for binding. We divide the clones of the "positive" pool (assuming just one) into two halves, check and repeat, etc. The desired clone, of the N in the library, is hence located in T tries, where $2^T \sim N$, or $T = \ln N / \ln 2 = \ln_2 N$. However, this requires the retention of records, has many sequential operations, must be redone for each subsequence to be detected, and is also error prone.

2. Nonadaptive one-pass pooling. An *experimental design* is chosen in which each pool contains predesignated clones, the whole set of pools is observed, and data are collected. No further experiments are carried out.

Example: Row and column design. $N = rc$ clones are placed at the cells of an r-row c-column grid; rows and columns are pooled separately. If the

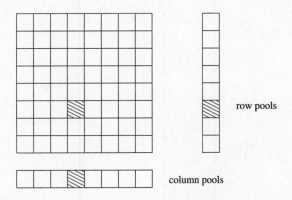

row pools

column pools

"word" is rare enough and the clone overlap is small enough, one column and one row will show positive, and so the word-containing clone is located. Optimally, to minimize the required number of pools, $r = c = N^{1/2}$, so there are $v = 2N^{1/2}$ pools. We can in principle use a D-dimensional hypercube design with $(D-1)$-dimensional slabs pooled, so $v = DN^{1/D}$, minimized at $D = \ln N$; then $v = e \ln N$ is quite small, but the setup is too complicated. Of course, when the word appears in several clones (we cannot have less than

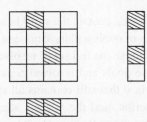

the coverage, on the average) there remain unresolved positions because only the *set* of rows and the *set* of columns are known. We reduce this ambiguity by applying the procedure to subsets of clones or by then examining separately each unresolved clone, partially but minimally adaptive.

What is an optimal "idiot-proof" design to reduce the pool number at a given level of resolution (number of unresolved clones)? This should presumably be an "on-the-average" strategy, as the identity of the word-carrying clones is to be regarded as unknown. We examine this issue mainly by replacing probability arguments by explicit Boolean operations, reducing the

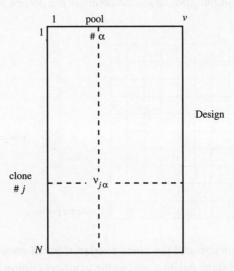

artistic and strengthening the mechanistic components of our discussion. We start in general by setting up an *incidence matrix* v to define our experimental design. Here

$$v_{j\alpha} = \begin{cases} 1 & \text{if clone } j \in \text{pool } \alpha \\ 0 & \text{if } j \notin \alpha \end{cases}.$$

Which clones are positive, i.e., contain the word being probed, is not known to us, but the vast majority of pools will have none of the few positive clones. These negative pools comprise the full set of resolved or definitely negative clones. When the negative pools and resolved negative clones are excised, we get a new small matrix $\bar{v}$ that still contains all the positive information. The simplest design criterion, and the one we adapt here, is to minimize the number of clones left over, as they are the "noise" in which the positive clones are embedded. (A presumably better criterion, but harder to design for, recognizes that some positive pools in $\bar{v}$ may contain just one remaining clone; such a clone is definite or resolved positive, and it makes sense to try to maximize their number – but we will not do so here.)

The way that probability enters the picture at this stage is that if the small target word occurs only once in the full fragment, but the library coverage is c, it would be expected to occur c times in the library. The best we can do is introduce a clone designator

$$\tau_i = \begin{cases} 1 & \text{if clone } i \text{ is } (+) \\ 0 & \text{if clone } i \text{ is } (-) \end{cases},$$

$$i = 1, \ldots N,$$

with the knowledge that c of the τ_i are 1 on the average, the rest being 0. Thus the average of any of the τ_i is c/N, and they are independent of each other. In other words, with $\langle\ \rangle$ denoting expectation,

$$\langle \tau_i \rangle = c/N \quad \equiv p.$$

The pair of sets $\{v_{i\alpha}, \tau_i\}$ of known/unknown quantities, 0 or 1, completely specifies the setup. Let us develop the required derived quantities. To start with, we define

$$Q_\alpha = \begin{cases} 1 & \text{if no clone in pool is } (+)\text{: } \alpha \text{ is negative} \\ 0 & \text{if at least one clone in } \alpha \text{ is } (+)\text{: } \alpha \text{ is positive} \end{cases}.$$

This is clearly given by the expression

$$Q_\alpha = \prod_{i=1}^{N} (1 - \tau_i \, v_{i\alpha}),$$

which is 1 if and only if every $\tau_i \, v_{i\alpha} = 0$, i.e., either $\{v_{i\alpha} = 0,\ i \text{ not in } \alpha\}$ or $\{v_{i\alpha} = 1 \text{ but } \tau_i = 0 : i \text{ is in } \alpha \text{ but } i \text{ is negative}\}$.

Next, clone i will be proved or resolved negative if it is a member of *some* negative pool, i.e., if $v_{i\alpha} = 1$ and $Q_\alpha = 1$ for at least one α:

$$i \text{ is resolved negative if}$$

$$A_{i\alpha} \equiv Q_\alpha v_{i\alpha} = 1 \text{ for at least one } \alpha.$$

This is equivalent to saying that

$$\prod_{\alpha=1}^{v}(1 - A_{i\alpha}) = 0.$$

Hence, if

$$P_i = 1 - \prod_{\alpha=1}^{v}(1 - A_{i\alpha}),$$

then

$$P_i = \begin{cases} 1 & \text{if } i \text{ is resolved negative} \\ 0 & \text{if } i \text{ is not resolved negative} \end{cases}.$$

Consequently, for the probability that i is resolved negative,

$$\text{Pr(clone } i \text{ is resolved negative)} = 1 - \left\langle \prod_{\alpha=1}^{v}(1 - A_{i\alpha}) \right\rangle.$$

As there is no a priori way of distinguishing the characteristics of two different clones, we might as well focus on clone #1:

$$\text{Pr(#1 is resolved } (-)) = 1 - \left\langle \prod_{\alpha=1}^{v}(1 - A_{\alpha}) \right\rangle$$

$$\text{where } A_{\alpha} = Q_{\alpha} v_{1\alpha} \quad \text{and } Q_{\alpha} = \prod_{i=1}^{N}(1 - \tau_i \, v_{i\alpha}).$$

A bit more conveniently, $\text{Pr(#1 is } (+)) = p$ means that $\text{Pr(#1 is } (-)) = 1 - p$; if we subtract the resolved $(-)$ clones from the merely $(-)$ clones, we get the actual but unresolved $(-)$ clones:

$$\text{Pr(#1 is unresolved negative)} = \left\langle \prod_{\alpha=1}^{v}(1 - A_{\alpha}) \right\rangle - p.$$

Carrying out this average requires expanding the product $\prod_1^v(1 - A_\alpha)$ and corresponds to the "inclusion–exclusion" theorem of probability; see, e.g., Percus (1971). The expansion consists, to within sign, of terms A_α, pair terms $A_\alpha A_\beta$, triplets $A_\alpha A_\beta A_\gamma, \ldots$, with the condition that no two indices are equal, and that permuting the order, e.g., $A_\alpha A_\beta A_\gamma \to A_\gamma A_\alpha A_\beta$, does not create a distinct term. We can take care of the latter by allowing everything, but dividing by $s!$, the number of permutations of s distinct letters. In other words

$$\prod_{\alpha=1}^{v}(1 - A_{\alpha}) = 1 - \sum_{\alpha=1}^{v} A_{\alpha} + \frac{1}{2!} \sum_{\alpha \neq \beta = 1}^{v} A_{\alpha} A_{\beta}$$

$$- \frac{1}{3!} \sum_{\alpha \neq \beta \neq \gamma = 1}^{v} A_{\alpha} A_{\beta} A_{\gamma} + \cdots,$$

and we conclude that

Pr(#1 is unresolved negative):

$$1 - p - \sum_{\alpha=1}^{v} \langle A_{\alpha} \rangle + \frac{1}{2!} \sum_{\alpha \neq \beta = 1}^{v} \langle A_{\alpha} A_{\beta} \rangle - \frac{1}{3!} \sum_{\alpha \neq \beta \neq \gamma = 1}^{v} \langle A_{\alpha} A_{\beta} A_{\gamma} \rangle \ldots.$$

For the evaluation, we see first that, because $v_{1\alpha}^2 = v_{1\alpha}$, then $A_\alpha = v_{1\alpha} \prod_{i=1}^N (1 - \tau_i v_{i\alpha}) = v_{1\alpha}(1 - \tau_1 v_1) \prod_{i=1}^N (1 - \tau_i v_{i\alpha})$, or

$$A_\alpha = (1 - \tau_1) v_{1\alpha} \prod_{i=2}^N (1 - \tau_i v_{i\alpha}).$$

However, $\langle \tau_1 \rangle = \cdots = \langle \tau_i \rangle = p$, so

$$\langle A_\alpha \rangle = (1 - p) v_{1\alpha} \prod_{i=2}^N (1 - p v_{i\alpha}).$$

Then, using $\tau_1^2 = \tau_1, \ldots \tau_i^2 = \tau_i$, we have

$$A_\alpha A_\beta = (1 - \tau_1)^2 v_{1\alpha} v_{1\beta} \prod_{i=2}^N (1 - \tau_i v_{i\alpha})(1 - \tau_i v_{i\beta})$$

$$= (1 - \tau_1) v_{1\alpha} v_{1\beta} \prod_{i=2}^N \{1 - \tau_i[1 - (1 - v_{i\alpha})(1 - v_{i\beta})]\},$$

and similarly

$$A_\alpha A_\beta A_\gamma$$
$$= (1 - \tau_1) v_{1\alpha} v_{1\beta} v_{1\gamma} \prod_{i=2}^N \{1 - \tau_i[1 - (1 - v_{i\alpha})(1 - v_{i\beta})(1 - v_{i\gamma})]\},$$

etc., and we conclude that, in general,

$$\langle A_\alpha A_\beta A_\gamma \ldots \rangle$$
$$= (1 - p) v_{1\alpha} v_{1\beta} v_{1\gamma} \ldots \prod_{i=2}^N [1 - p + p(1 - v_{i\alpha})(1 - v_{i\beta})(1 - v_{i\gamma}) \ldots].$$

We therefore have the relatively simple result that, *at fixed design*,

$$\frac{1}{1 - p} \Pr(\#1 \text{ is unresolved negative})$$

$$= 1 - \sum_\alpha \left\{ v_{1\alpha} \prod_2^N [1 - p + p(1 - v_{i\alpha})] \right\}$$

$$+ \frac{1}{2!} \sum_{\alpha \neq \beta} \left\{ v_{1\alpha} v_{1\beta} \prod_2^N [1 - p + p(1 - v_{i\alpha})(1 - v_{i\beta})] \right\}$$

$$- \frac{1}{3!} \sum_{\alpha \neq \beta \neq \gamma} \left\{ v_{1\alpha} v_{1\beta} v_{1\gamma} \prod_2^N [1 - p + p(1 - v_{i\alpha})(1 - v_{i\beta})(1 - v_{i\gamma})] \right\}$$

$$+ \cdots. \tag{2.1}$$

Now how in the world do we use Eq. (2.1) to, e.g., plan a design that minimizes the number of pools required for achieving a given resolution? Any complicated schedule would defeat the whole objective. A neat way is to hedge our bets and choose the design in some random fashion. The simplest is random incidence: Just choose a parameter k so that each clone occurs on the average in k pools. Hence $\langle \sum_{\alpha=1}^{v} v_{i\alpha} \rangle = k$ for any i, or because the $v_{i\alpha}$ for different α are independent,

$$\langle v_{i\alpha} \rangle = k/v, \quad \text{independently.}$$

The averaging of Eq. (2.1) is then next to trivial: The sum $\sum_{\alpha_1 \neq \alpha_2 ... \neq \alpha_s}$ $\langle A_{\alpha_1} \cdots A_{\alpha_2} \rangle$ contains $v(v-1)\cdots(v-(s-1))$ terms, all of which have the same value $v(v-1)-(v-(s-1))/s! = v!/s!v-s!$, the binomial coefficient $\binom{v}{s}$, and every v is replaced by k/v in the average. We thus have

$$\frac{1}{1-p} \Pr(\#1 \text{ is unresolved negative})$$

$$= 1 - \binom{v}{1}\left(\frac{k}{v}\right)\left[1 - p + p\left(1 - \frac{k}{v}\right)\right]^{N-1}$$

$$+ \binom{v}{2}\left(\frac{k}{v}\right)^2\left[1 - p + p\left(1 - \frac{k}{v}\right)^2\right]^{N-1}$$

$$- \binom{v}{3}\left(\frac{k}{v}\right)^3\left[1 - p + p\left(1 - \frac{k}{v}\right)^3\right]^{N-1} + \cdots.$$

Also, N is large, but p is small, and $p[1 - (1 - \frac{k}{v})^s]$ even smaller, so we can use $(1-x)^{N-1} \to e^{-Nx}$. Finally, because $1 - p = \Pr(\#1 \text{ is negative})$, we can write

$$P_{\text{indef}}^{-} \equiv \Pr[\#1 \text{ is unresolved } (-) \mid \#1 \text{ is } (-)]$$

$$= \frac{1}{1-p} \Pr(\#1 \text{ is unresolved } (-))$$

$$= \sum_{t=0}^{v} (-1)^t \binom{v}{t}\left(\frac{k}{v}\right)^t \exp -c\left[1 - \left(1 - \frac{k}{v}\right)^t\right], \quad (2.2)$$

where $Np = c$ is just the coverage.

This is a nice explicit formula, but it has v terms, and they almost cancel [if the exponential were constant, we would have $\sum_{t=0}^{v}(-1)^t\binom{v}{t} = (1-1)^v = 0$]. Therefore numerical evaluation and tabulation are problems. We prefer that numerical evaluation be at the conclusion of analytical processing, so that, at the very least, we can get a better feeling for parametric dependence. A number of analytic tricks are available. Here is one: We introduce the asymmetric difference operator

$$\Delta_s^\lambda f(s) = \lambda f(s+1) - f(s) \equiv (\lambda E - 1) f(s),$$

and hence

$$\left(\Delta_s^\lambda\right)^v f(s) = (\lambda E - 1)^v f(s) = \sum_0^v (-1)^{v+t} \binom{v}{t} \lambda^t f(s+t),$$

so that in fact

$$P_{\text{indef}}^- = e^{-c}(-1)^t \left(\Delta_s^{k/v}\right)^v e^{c(1-\frac{k}{v})^s}|_{s=0}$$

$$= e^{-c} \left(-\Delta_s^{k/v}\right)^v \sum c^t/t! \left(1-\frac{k}{v}\right)^{st}\bigg|_{s=0}.$$

However,

$$-\Delta_s^{k/v} \left(1-\frac{k}{v}\right)^{st} = \left(1-\frac{k}{v}\right)^{st} - \frac{k}{v}\left(1-\frac{k}{v}\right)^{(s+1)t}$$

$$= \left[1 - \frac{k}{v}\left(1-\frac{k}{v}\right)^t\right]\left(1-\frac{k}{v}\right)^{st},$$

so

$$P_{\text{indef}} = e^{-c} \sum \frac{c^t}{t!}\left[1 - \frac{k}{v}\left(1-\frac{k}{v}\right)^t\right]^v$$

$$= \sum_t \left(1 - \frac{k}{v}e^{-\gamma t}\right)^v \frac{c^t}{t!}e^{-t} \sim \sum_t e^{-k\,e^{\gamma t}}\frac{c^t}{t!}e^{-t}, \qquad (2.3)$$

where

$$\gamma = -\ln\left(1-\frac{k}{v}\right).$$

Now all terms are positive, so computation is quick and accurate.

In practice, a restricted random-incidence design, in which we impose a fixed number of pools entered by each clone, $\sum_\alpha v_{i\alpha} = k$ for each i, has experimental advantages. It can be treated similarly, but the analysis is a bit more involved.

Assignment 3

1. Choose reasonable values of c, k, and v in Eq. (2.2) and carry out the evaluation at a few levels of precision to see how much computational accuracy is needed.

2. Because Eq. (2.3) has only positive terms, analytic approximations are controllable. Try one or two and compare with problem 1.

2.5. Reprise

There are, as we have seen, various ways of joining clones and building up islands for the purpose of constructing maps of the genome, and more are being suggested all the time. A dominant role in the increasing common whole genome "shotgun sequencing" is now being played by the use of end-characterized clones, i.e., those in which a few hundred base pairs at each end are identified, base after base, by use of the old Sanger technique (Sanger et al., 1977). Then overlap is recognized if the leading end of a "new" clone overlaps, by at least T base pairs, the trailing end of the last clone of the currently developed island. Rather than develop a new clever technique of analysis each time, it might pay to create a descriptive machinery that is tailored to the general concept and therefore allows diverse questions to be answered more routinely. This is not a novel concept, but let us carry it a bit further than is usually done (Percus and Percus, 1999). We deal here with only the basic fingerprint assembly at fixed clone length L and overlap criterion T, but in a fashion that extension to a distribution of clone size and overlap threshold can be carried out with relative ease.

If genome end effects are unimportant, we can imagine that the island-building process starts with a clone whose left end is at site 1, its right end at its length L. Now we build up an island by putting down a clone at step k, starting at site k, with probability $p(= N/G)$. If it overlaps the current $(k-1)$ island sufficiently, a new island with right end at $R_k = k + L - 1$ will be produced; if it does not, the island terminates at the previous value R_{k-1}. If no clone is found starting at k, with probability $q = 1 - p$, then $R_k = R_{k-1}$.

The first crucial criterion is that of sufficient overlap, T. In the basic case we are considering, this simply requires that a clone be found starting at site k such that

$$R_{k-1} \geq k + T - 1.$$

The complementary criterion is that of the island terminating; it will do so at step $k - 1$ as soon as $R_{k-1} = k + T - 2$, for then a clone starting at k can have an overlap of at most $T - 1$. Hence, at step k,

$$R_k = k + T - 1 : \text{ stop.}$$

Pictorially, then, the island develops as a random walk in which at each step, $\Delta k = 1$, either a clone is added, so that R_k jumps to $k + L - 1$, or no clone is added, so that $R_k = R_{k-1}$, but the island stops if this results in $R_k = k + T - 1$.

What we want to do now is to find

$$P_k(R),$$

which is the probability that our developing island at step k has length R.

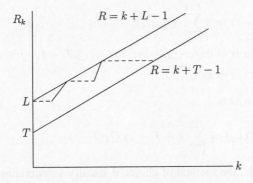

Because an island is completed when $R_k = k + T - 1$, the expected length of an island will be just

$$E(L_I) = \sum_{k=1}^{\infty} (k + T - 1) \; P_k(k + T - 1).$$

$P_k(R)$ is to be found iteratively. If the island has "arrived" at the point (k, R), it can have done so in two ways. Either (1) $R = k + L - 1$, corresponding to an overlapping clone appearing at step k, and this has probability p providing that R_{k-1} was anything between its minimum $(k - 1) + T$ and its maximum $(k - 1) + (L - 1)$,

$$P_k(k + L - 1) = p \sum_{R'=k-1+T}^{k-2+L} P_{k-1}(R'),$$

or (2) $R = R_{k-1}$ because no clone appears at step k, and of course R_{k-1} cannot have been smaller than $(k - 1) + T$ (or the island would already have terminated):

$$P_k(R) = q \, P_{k-1}(R) \quad \text{if} \quad R \geq k - 1 + T, \quad R \leq k - 2 + L.$$

Finally, the whole process starts at

$$P_1(R) = \delta_{R,L}$$

(the Kronecker delta $\delta_{i,j} = 1$ if $i = j$, otherwise 0).

The "walk" that we are analyzing is one with a reflecting barrier along the slope $R = k + L - 1$, an absorbing barrier along $R = k + T - 1$. Analysis is much easier if the barriers are fixed, so we set

$$P_k(k + j) = Q_k(j),$$

and thus have

$$Q_k(L - 1) = p \sum_{i=T}^{L-1} Q_{k-1}(i),$$

$$Q_k(j) = q \, Q_{k-1}(j + 1) \quad \text{for} \quad j \geq T - 1, \quad j \leq L - 2,$$

$$Q_1(j) = \delta_{j,L-1},$$

subsequent to which

$$E(L_I) = \sum_{k=1}^{\infty} (k + T - 1) \, Q_k(T - 1).$$

For solving this, the method of choice is usually a generating function. We set

$$Q(x, \, j) = \sum_{k=1}^{\infty} x^k \, Q_k(j)$$

$$= x \, \delta_{j,L-1} + \sum_{k=2}^{\infty} x^k \, Q_k(j).$$

Now we multiply the recursion relations by x^k and sum over k from 2 to ∞. We see that

$$Q(x, \, L - 1) = x + p \sum_{i=T}^{L-1} x \, Q(x, \, i),$$

$$Q(x, \, j) = q \, x \, Q(x, \, j + 1) \quad \text{for} \quad L - 2 \geq j \geq T - 1,$$

$$E(L_I) = \partial Q(x, \, T - 1)/\partial x|_{x=1} + (T - 1) \, Q(1, \, T - 1).$$

Working up from $Q(x, \, T - 1)$ in the second relation, we have $Q(x, \, j) = Q(x, \, T - 1)/(qx)^{j+1-T}$ for $j \geq T - 1$, and substituting this into the first relation, we obtain

$$\frac{Q(x,\ T-1)}{(qx)^{L-T}} = x + px\ Q(x,\ T-1) \sum_{i=T}^{L-1} \frac{1}{(qx)^{i+1-T}}$$

$$= x + \frac{p}{q}\ Q(x,\ T-1) \left[\left(\frac{1}{qx}\right)^{L-T} - 1 \right] \bigg/ \left(\frac{1}{qx} - 1\right),$$

so that

$$Q(x,\ T-1) = [x(1-qx)(qx)^{L-T}]/[1-qx-px(1-qx)^{L-T})]$$

$$= [x(1-qx)(qx)^{L-T}]/[1-x+px(qx)^{L-T}].$$

We see that $Q(1,\ T-1) = 1$, which just says that the probability that the island ends someplace is 1. What we need then is $Q'_x(1,\ T-1)$, and a brief calculation shows that

$$Q'_x(1,\ T-1) = \frac{1}{p} \left[q^{-(L-T)} - q \right],$$

from which we conclude that

$$E(L_I) = T + \frac{1}{p} \left[q^{-(L-T)} - 1 \right],$$

or, because $q^{-(L-T)} = (1-p)^{-(L-T)} = e^{p(L-T)}$ for small p, that

$$E(L_I) = T + \frac{1}{p} \left[e^{p(L-T)} - 1 \right],$$

which we have already seen.

What deeper questions can we ask? The simplest might be about the dispersion, or standard deviation, of L_I (after all, a mean of 10 is not too germane if 10% of the population are 100 and the rest are 0):

$$\sigma(L_I) = \left[E\left(L_I^2\right) - E(L_I)^2 \right]^{1/2}$$

$$= \left\{ \sum_1^\infty k^2\ Q_k(T-1) - \left[\sum_1^\infty k\ Q_k(T-1) \right]^2 \right\}^{1/2}$$

(the constant part of L_I does not contribute)

$$= [Q''_x(1) + Q'_x(1) - Q'_x(1)^2]^{1/2}$$

$$= [(\ln Q_x)''(1) + (\ln Q_x)'(1)]^{1/2},$$

where the argument $T - 1$ is omitted. After some minor algebra,

$$\sigma(L_I) = \left\{ \left[T + \frac{q^2}{p^2} - \frac{2L}{pq^{L-T}} + \frac{1}{p^2 q^{2(L-T)}} \right] + \left(-T - \frac{q}{p} + \frac{1}{pq^{L-T}} \right) \right\}^{1/2}$$

$$= \left[\frac{q^2}{p^2} - \frac{q}{p} - \frac{2L-1}{pq^{L-T}} + \frac{1}{p^2 q^{2(L-T)}} \right]^{1/2}$$

$$\rightarrow \frac{1}{p} e^{p(L-T)} \left[1 - \left(L - \frac{1}{2} \right) p e^{-p(L-T)} \cdots \right],$$

in fact very close to $E(L_I)$ itself, as in an exponential, or basic survival time, distribution – which this is very close to.

It is a bit more complicated to restrict the islands to contigs, which are the objects would really pick up. This means omitting those islands that have not hit the upper barrier, i.e., that have not encountered at least one weight-p jump. For this purpose, we simply refrain from using the fact that $p + q = 1$ in our first expression for $Q(x, T - 1)$:

$$Q_p(x, T - 1) = q^{L-T} x^{L-T} (1 - qx)/[1 - (p + q)x + px q^{L-T} x^{L-T}].$$

The coefficient of p^k then represents the weight of the $(k + 1)$-clone islands. Subtracting out $Q_0(x, T - 1)$ gives the contig-generating function

$$\bar{Q}_{con}(x, T - 1) = Q(x, T - 1) - Q_0(x, T - 1)$$

$$= q^{L-T} x^{L-T} px^2 [1 - (qx)^{L-T}]/[1 - x + px(qx)^{L-T}],$$

which is no longer normalized: $\bar{Q}_{con}(1, T - 1) = 1 - q^{L-T}$, so that

$$Q_{con}(x, T - 1) = \frac{q^{L-T}}{1 - q^{L-T}} px^{L+2-T} \frac{1 - (qx)^{L-T}}{1 - x + px(qx)^{L-T}}.$$

Hence

$$Q'_{con}(1, T - 1) = 1 + \frac{1}{p} q^{-(L-T)} - (L - T) \frac{q^{L-T}}{1 - q^{L-T}},$$

$$E(L_{I, con}) + \frac{1}{p} q^{-(L-T)} + \frac{T - L q^{L-T}}{1 - q^{L-T}}.$$

The advantage of this mode of derivation is that generalization to a distribution of clone lengths, $f(L)$, and a distribution of overlap detection thresholds, $w(T)$, is easy to carry out. Although a complete closed-form analysis is available in only special cases, means and variances of required characteristics can be obtained quite generally. For example, for the mean island length, we find

$$E(L_I) = \sum_0^\infty \bar{F}(R) \Big/ \prod_0^R q(r),$$

where

$$\bar{F}(R) = \sum_{R+1}^\infty f(L), \quad q(r) = 1 - pw(r)\bar{F}(r).$$

In the continuum limit, in which the unit length, say T_0, corresponds to very many base pairs, writing

$$f_c(l) = \lim_{T_0 \to \infty} T_0\, f(l\, T_0), \quad W(r) = w(r/T_0), \quad \rho = T_0\, p,$$

we see that this reduces to

$$E(l_I) = \int \bar{F}_c(l) \exp\left(\rho \int_0^l \bar{F}_c(t)\, W(t)\, dt\right) dl,$$

showing that the generalization in Section 2.1 from the fixed clone length, $\bar{F}(R) = \theta(L - R)$, fixed threshold, $w(R) = \theta(R - T)$, case was a bit naive.

3

Sequence Statistics

Once we can assume that long stretches of DNA are completely sequenced, it is possible to analyze the information contained therein. In the "language" of DNA, there are many "verbal" biases – dialects, accents, pauses – that have evolved for broad reasons of physical and biochemical accessibility and function. They contribute to the "default state," the random or null hypothesis with respect to which additional information must be assessed. Ignorance of this background bias not only ignores available information but also poses as noise against which a signal must compete. In broad outline, as we have noted, we know that DNA is not homogeneous, but contains regions that code for specific proteins as well as those with associated regulatory functions. Each of the former is divided into *exons*, which (three bases at a time) are transcribed and translated into amino acids, *introns*, which are spliced out during this process, and instructional subsequences. The latter are often confined to a *flanking* region of this coding section, all residing in a sea of "junk DNA," which may or may not be functional.

3.1. Local Properties of DNA

To start with, the bases are not even at equal frequencies. For example, for the human mitochondrion (17,000 bp), we have typically [see, e.g., Weir (1990), Chap. VII]

A	C	G	T
0.31	0.31	0.25	0.13

For different regions of the human fetal globin gene, the pair G_γ, A_γ, the distributions are given in the following table.

42

	Total Length	A	C	G	T
5' flanking (2)	1000	0.33	0.23	0.22	0.22
3' flanking (2)	1000	0.29	0.15	0.26	0.30
Introns (4)	1996	0.27	0.17	0.27	0.29
Exons (6)	882	0.24	0.25	0.28	0.22
Intergenic (1)	2487	0.32	0.19	0.18	0.31

Clearly, this bias should be taken into account, and we should bunch together, for statistical purposes, only subsequences of similar character. However, even if this is not done (choosing 140,000 bp from 166 vertebrate sequences as an example), an additional local structure immediately appears: If p_i is the relative frequency of base i, p_{ij} of the pair $5' \cdots ij \cdots 3'$, then the pair correlation ratio

$$p_{ij} / p_i p_j$$

is not at all unity, as it would be for independent placement, but rather is as in the following table.

		A	C	G	T	Second Base
	A	1.15	0.84	1.16	0.85	
First	C	1.15	1.18	0.42	1.26	
Base	G	1.04	0.99	1.14	0.82	
	T	0.65	1.00	1.29	1.07	

Observe the very small CG frequency, which presumably is due to geometric "mismatch."

Bunching together sequences of different singlet frequencies will masquerade as pair correlation, e.g., $\left(\begin{smallmatrix} 1 & 3 \\ 3 & 9 \end{smallmatrix}\right) + \left(\begin{smallmatrix} 9 & 3 \\ 3 & 1 \end{smallmatrix}\right) = \left(\begin{smallmatrix} 10 & 6 \\ 6 & 10 \end{smallmatrix}\right)$ does not have the form $N_\alpha N_\beta$ when the full sequence consists of two internally uncorrelated sequences; hence the tacit assumption of homogeneity is implicit in assessing correlations.

A fairer test for pair correlations is to stick to a single functional entity, here a chicken β-globin gene exon, organized again according to the number of successive pairs of sequences of the 16 possible types.

	A	C	G	T	Total
A	23	26	23	15	87
C	37	51	14	41	143
G	25	38	36	19	118
T	2	29	44	14	87
Total	87	144	117	89	437

Note that the N_{ij} in this *contingency table*, the actual number of times ij is found, do not quite satisfy $N_{i\cdot} = N_{\cdot i}$, etc., where $N_{i\cdot} \equiv \sum_j N_{ij}$, $N_{\cdot i} \equiv \sum_j N_{ji}$, because the left base of the leftmost pair of the chain cannot be the right base of any pair, etc.

The standard test for independent model probabilities is the χ^2 *test* (Cramer, 1946). This goes as follows: Suppose the result of a measurement is indexed by α, with N_α the number of times α occurs, and of course $\sum_{\alpha=1}^s N_\alpha = N$, for s types of result, is the total number of trials. If the probability of α is p_α, then, on the assumption of independence, the set $\{N_\alpha\}$ occurs with a probability

$$\left(N! \bigg/ \prod_{\alpha=1}^s N_\alpha! \right) \prod_{\alpha=1}^s p_\alpha^{N_\alpha}.$$

If N is large, we let $N_\alpha = N x_\alpha$, which converts the above to

$$P\{N_\alpha\} = \left[N! \bigg/ \prod_{\alpha=1}^s (N x_\alpha)! \right] \prod_1^s p_\alpha^{N x_\alpha},$$

and we then use Stirling's formula $n! \sim \sqrt{2\pi n}(n/e)^n$ to obtain

$$P\{N_\alpha\} \sim \frac{N!}{(2\pi N)^{s/2} N^N} \left[\prod_1^s \left(\frac{p_\alpha e}{x_\alpha} \right)^{x_\alpha} \right]^N \bigg/ \left(\prod_1^s x_\alpha \right)^{1/2}.$$

Now $x_\alpha \ln (p_\alpha e / x_\alpha)$ has a single maximum at $x_\alpha = p_\alpha$:

$$x_\alpha \ln (p_\alpha e / x_\alpha) = p_\alpha - \frac{1}{2p_\alpha} (x_\alpha - p_\alpha)^2 \cdots$$

so that the Nth power in $P\{N_\alpha\}$ has a very sharp maximum, and

$$P\{N_\alpha\} \sim \frac{N!/N^N e^N}{(\prod 2\pi N \, p_\alpha)^{1/2}} \, e^{-(N/2) \sum_1^s (x_\alpha - p_\alpha)^2 / p_\alpha},$$

depending on only the quantity

$$\chi^2 = N \sum_{\alpha=1}^s \frac{(x_\alpha - p_\alpha)^2}{p_\alpha} = \sum_\alpha \frac{(N_\alpha - p_\alpha N)^2}{p_\alpha N}$$

$$= \sum_\alpha \frac{(N_\alpha - \langle N_\alpha \rangle)^2}{\langle N_\alpha \rangle},$$

where we have used (and will use) the notation $\langle N_\alpha \rangle$ interchangeably with

$E(N_\alpha)$. Introducing the new variable

$$y_\alpha = N^{1/2} \frac{x_\alpha - p_\alpha}{p_\alpha^{1/2}},$$

we see that the distribution is independent of the p_α. For large N, we may regard the y_α as continuous versions of the N_α, as under a change $\Delta N_\alpha = 1$, we have $\Delta y_\alpha = N^{1/2} p_\alpha^{-1/2} \Delta x_\alpha = (Np_\alpha)^{-1/2} \Delta N_\alpha = (Np_\alpha)^{-1/2}$. If the $\{y_\alpha\}$ were independent, the probability for the $\{y_\alpha\}$ would then be determined by $p\{y_\alpha\} \Delta y^s = P\{N_\alpha\}(\Delta N_\alpha)^s$, or

$$p\{y_\alpha\} = N!/(N/e)^N \left(\frac{1}{2\pi}\right)^{s/2} e^{-(1/2)\chi^2}$$

where

$$\chi^2 = \sum_1^s y_\alpha^2.$$

The reason for the large prefactor $N!/(N/e)^N \sim \sqrt{2\pi N}$ will appear in a moment.

Now what is the distribution of χ^2? The y's are not free in s-dimensional y space because they are restricted to the hyperplane $\sum N_\alpha = N$, or

$$\sum_1^s p_\alpha^{1/2} y_\alpha = 0,$$

and it is the surface "thickness" of only $\sim N^{-1/2}$ on the restricted lattice of the $\{y_\alpha\}$ that is responsible for the above $\sqrt{2\pi N}$. However, we do not have to worry about normalization; it can be supplied as the last step. In general, suppose there are r linear homogeneous relations that the $N_\alpha - \langle N_\alpha \rangle$, and hence the y_α, have to satisfy. Then the space is reduced to dimensionality

$$\nu = s - r,$$

the number of *degrees of freedom*, but the intersection of the spherically symmetric unconstrained distribution with the hyperplane through the origin of restrictions remains spherically symmetric Gaussian on the ν-dimensional space. Call the new coordinates in this space the $\{Y_\alpha\}$. Then we seek the distribution of

$$\chi^2 = \sum_1^\nu Y_\alpha^2$$

where

$$p\{Y_\alpha\} = \prod_{\alpha=1}^{\nu} \left\{ \frac{1}{\sqrt{2\pi}} e^{[-(1/2)Y_\alpha^2]} \right\}$$

(the constant is now obtained by the normalization condition $\int \cdots \int p\{Y_\alpha\}$ $dY^\nu = 1$). Because

$$\langle e^{-\gamma\chi^2} \rangle = \prod_\alpha \int \frac{1}{\sqrt{2\pi}} e^{-(\gamma+1/2)Y_\alpha^2} dY_\alpha$$
$$= (1+2\gamma)^{-\nu/2}$$

defines $\int_0^\infty p(\chi^2) e^{-\gamma\chi^2} d\chi^2$, we readily find, on consulting inverse Laplace transform tables, that the χ^2 distribution for ν degrees of freedom is

$$p_\nu(\chi^2) = (\chi^2)^{\nu/2-1} e^{-(1/2)\chi^2} / 2^{\nu/2} \left(\frac{\nu}{2} - 1 \right)!,$$

from which it follows that, for any sizable ν, $p_\nu(\chi^2)$ is also sharply distributed, with $E(\chi^2) = \nu$.

Now back to our example. We want to test for independent juxtaposition, i.e., whether (noting the distinction between a left member of a pair and a right member, because of end effects) the observed $\{N_{ij}\}$ are consistent with the p_{ij}, computed as

$$p_{ij} = p_i. \, p_{.j}.$$

We do not actually know the $p_i.$ and $p_{.j}$; we have to estimate them by means of $p_i. = 1/N \sum_j N_{ij} = N_i./N$, $p_{.j} = N_{.j}/N$, and then the question is whether

$$\langle N_{ij} \rangle = N_i.N_{.j}/N$$

is a valid estimate, modified of course by known restrictions. However, there are many restrictions, because $\sum_j (N_{ij} - \frac{N_i.N_{.j}}{N}) = \sum_i (N_{ij} - \frac{N_i.N_{.j}}{N}) = 0$, as is readily verified, amounting to $[2 \times 4] - 1 = 7$ independent restrictions. We compute

$$\chi^2 = \sum_{i,j} \left(N_{ij} - \frac{N_i.N_{.j}}{N} \right)^2 \Big/ (N_i.N_{.j}/N) \sim 59.3$$

for the example quoted, compared with the $E(\chi^2) = 16 - 7 = 9$ degrees of freedom. This shows that the assumption $p_{ij} = p_i.p_{.j}$ is extremely inconsistent with the data.

However, is the strictly pair-associating Markov chain assumption good enough? Maybe there are really three, or four, or more successive intrinsic base correlations. Let us use the notation

$$\delta_x(j) = \begin{cases} 1, \text{ base } j \text{ is at site } x \\ 0, \text{ base } i \neq j \text{ is at site } x \end{cases}.$$

Then the multisite correlations are defined as

$$p_i = Av_x\, \delta_x(i),$$
$$p_{ij} = Av_x\, \delta_x(i)\,\delta_{x+1}(j),$$
$$p_{ijk} = Av_x\, \delta_x(i)\,\delta_{x+1}(j)\,\delta_{x+2}(k)\,,\ldots,$$

where Av_x means moving the beginning of the multiplet over all possible chain positions and averaging. The corresponding conditional correlations are then defined as

$$p_{j|i} = p_{ij}/p_i, \qquad p_{k|ij} = p_{ijk}/p_{ij}\,,\ldots;$$

for independence, a Markov chain of order 0, $p_{j|i} = p_j$, or $p_{ijk} = p_i p_j p_k \cdots$;

for a Markov chain of order 1, $p_{k|ij} = p_{k|j}$, or $p_{ijkl\cdots} = \frac{p_{ij}p_{jk}p_{kl}\cdots}{p_j p_k p_l \cdots}$;

for a Markov chain of order K, $p_{j|i_1\cdots i_L} = p_{j|i_{L-K+1}\cdots i_L}$ for $L \geq K$, or

$$p_{i_1 i_2 \cdots} = \frac{p_{i_1 \cdots i_{K+1}} p_{i_2 \cdots i_{K+2}} p_{i_3 \cdots i_{K+3}} \cdots}{p_{i_2 \cdots i_{K+1}}\, p_{i_3 \cdots p_{K+2}} \cdots}.$$

Because $p_{i_2\cdots i_{K+1}} = \sum_{i_1} p_{i_1\cdots i_{K+1}}$, the Kth-order chain is fully specified by $4^{K+1} - 1$ independent parameters, and the crucial $p_{j|i_1\cdots i_K}$ by 3×4^K (as $\sum_j p_j = 1$). Clearly we need a lot of data – i.e., very long chains – to determine these parameters, *assuming* homogeneity of the chain.

Perhaps we can at least get the intrinsic correlation length in the sense of the order of K of the underlying Markov chain (Katz, 1981); one technique is (Tong, 1975; Tavare and Giddings, 1989) the Bayesian information criterion (BIC). Consider the subsequence $i_x i_{x+1} \cdots i_{x+K'}$ on the test sequence as the (x, K') piece of data, data(x, K'), and choose some subset of the $\{(x, K')\}$. Given this data, then, according to the familiar compound probability theorem, the a posteriori probability that a given K is the correct order is

$$P[K \mid \{\text{data}(x, K')\}] = P[\{\text{data}(x, K')\}\mid K]\, P(K)/P[\{\text{data}(x, K')\}].$$

Hence we want to choose K to maximize the *likelihood* $P[\{\text{data}(x, K')\}\mid K] = \prod_{(x,K)} P[\text{data}(x, K) \mid K]$, evaluated as

$$\prod_{(j_1,\ldots,j_{K+1})} P\,(j_{K+1} \mid j_1 \cdots j_K; K)^{n(j_1,\ldots,j_{K+1})},$$

where $n(j_1, \ldots, j_{K+1})$ is the number of times the subsequence $j_1 \ldots j_{K+1}$ is encountered in the data. In other words, we want to maximize

$$\ln P[\{\text{data}(x, K')\} \mid K] = \sum_{j_1 \cdots j_{K+1}} n(j_1 \ldots j_{K+1}) \ln \frac{n(j_1 \ldots j_{K+1})}{\sum_j n(j_1 \ldots j_K j)}.$$

However, there are 3×4^K parameters that are implicitly determined in the process, each as the result of $n = \sum_{j_1 \cdots j_{K+1}} n(j_1 \cdots j_{K+1})$ pieces of data. This suggests a parametric-overkill penalty function, and K_{BIC} is defined as the K that maximizes

$$B(K) = \sum_{j_1 \cdots j_{K+1}} n(j_1 \cdots j_{K+1}) \ln \frac{n(j_1 \cdots j_{K+1})}{\sum_j n(j_1 \cdots j_K j)} - \frac{1}{2} 3 \times 4^K \ln n.$$

The penalty function is not unique, satisfying here weak information-theoretic criteria, but we can indeed show that

$$\lim_{n \to \infty} Pr(K_{\text{BIC}} = K_{\text{true}}) = 1.$$

Example: The 48,500-bp "head region" of phage λ. We have

K	0	1	2	3	4
$-2B(K)$	23,532	23,352	23,113	24,114	28,434

showing (weakly) that $K = 2$ is suggested, consistent at least with the triplet structure of $DNA \to$ amino acid translation.

Once an estimate of K is made, we can return to estimate the parameters $p_{i|j_1 \cdots j_K}$. Because there are far too many for a reasonable statistical estimation on a guaranteed homogeneous piece of DNA, one way out is to model the parameter set (see, e.g., Raftery, 1985)

$$p_{i|j_1 \cdots j_K} = \sum_{l=1}^{K} \lambda_l q_{i|j_l},$$

where

$$\sum_l \lambda_l = 1 \quad (K - 1 \text{ parameters}),$$

$$\sum_i q_{i|j} = 1 \quad (12 \text{ parameters}),$$

a total of only $K + 11$ parameters. These can then be found from the same log-likelihood estimator, i.e.,

$$L = \sum n(j_1 \cdots j_K i) \ln \sum_{l=1}^{k} \lambda_l q_{i|j_l},$$

subject to

$$\sum_1^k \lambda_l = 1, \qquad \sum_i q_{i|j} = 1,$$

without the necessity of an additional penalty function.

3.2. Long-Range Properties of DNA

Regions of uniform A, C, G, and T distribution at low resolution are called *isochores* (Bernardi, 1989); they can stretch from kilobase pairs (Kb) to hundreds of kilobase pairs. Because DNA data as a whole consist of double strands, distributions are often specified by the $G + C$ content of either or both strands, which shifts when the isochore switches. One way (Román-Roldán, 1998) of detecting an isochore or other segmented structure is (Bernaola-Galván et al., 2000) by means of the Jensen–Shannon entropy. If the sequence S is decomposed into segments S_i, l_i long and with base frequencies $p_{i\alpha}$, then the S_i entropy per site is defined as $H(S_i) = -\sum_\alpha p_{i\alpha} \ln_2 p_{i\alpha}$, and the total Jensen–Shannon entropy is

$$JS\{S_i\} = \sum_i (l_i/L)[H(S) - H(S_i)] \ge 0,$$

where

$$L = \sum_{i=1}^n l_i.$$

For each decomposition of an intelligently refined set of likely decompositions, we find the probability P that a random decomposition of S has a value equal to or less than $JS\{S_i\}$ and takes the maximizing decomposition $\{S_i\}$ at fixed P as the optimal decomposition at confidence level P. Other statistical techniques have been suggested (Ramensky et al., 2000).

If this structure is believed literally, it must be detected and taken into account. However, are there correlations within such long stretches, between successive ones, and how do we measure them? These are questions concerning the long-range properties of DNA.

3.2.1. Longest Repeat

A rather general criterion of obvious biological relevance – typical of a number of tests used by Karlin and collaborators (see, e.g., Karlin et al. 1989) but not directly expressible in terms of correlations, is as follows. For a very long sequence, any subsequence may be expected to repeat, e.g., a seven-base subsequence every $4^7 \sim 16,000$ bp, but longer ones may perhaps not repeat at all. A repeat of a longer subsequence might indicate biological function, evolutionary history, etc., and indeed be significant. The most concise way of checking would be to look at the longest subsequence that repeats and ask whether this is a reasonable consequence of the length of the full sequence. For that purpose, a good thing to do would be to compute the mean and the variance of the longest repeat length, e.g., for simplicity on the assumption of independently distributed bases, and compare with observation. Therefore let us consider a chain of length n in base pairs. To search for subsequence repeats (computationally, we may first amass low-order repeats, and successively refine them by raising the threshold order), we construct the self-comparison "dot matrix"

$$A_{xy} = \begin{cases} 1, & \text{if } x \neq y \text{ but } j_x = j_y \\ 0, & \text{otherwise} \end{cases},$$

where base type is denoted by j, base location by $x, y, \ldots$;

```
        A   T   T   G   A   T
    A               •
                          ⋱
    T       •                   •
    T   •                       •
    G
    A •
              ⋱
    T           •       •   •
```

here, 1s in the matrix are represented by dots, 0s by no entry. A run of r dots down some upper or lower diagonal (i.e., at $-45°$) then signifies a repeat of an r subsequence (AT in the example shown). This matrix is symmetric. If we run half the diagonals – say the upper ones – one after the other, creating a string of $n(n-1)/2$ sites, we will include only a negligible fraction r/n of sites within r of the end of a diagonal that are recorded as possible members of r runs, but should not be. Also, although the $n \times n$ matrix represents only n pieces of information, the probability of finding a repeat at the longest repeating subsequence is very small to start with, so we do not have to worry about correlations between run locations.

Hence the probability that the longest repeat is of length r is the same as the probability P_l (r = max) that the longest run of "successes," or dots on a chain of length $l \sim \frac{1}{2} n^2$, is r. A success means the matching of two bases and hence occurs with probability

$$p = \sum_1^4 p_i^2.$$

Now P_l (r = max) in turn is simply the probability that there is no $r + 1$ run minus the probability that there is no r run:

$$P_l (r = \text{max}) = \bar{P}_l(r + 1) - \bar{P}_l(r)$$

over a sequence of length $l \cong \frac{1}{2} n^2$. [Technically, $P(r = \text{max}) = P(r$ run and no $r + 1$ run$) = 1 - P($no r run, or an $r + 1$ run$) = 1 - \bar{P}(r$ run$) - P(r + 1$ run$) - P($no r run but an $r + 1$ run$) = 1 - \bar{P}(r) - P(r + 1) - 0.$] Let $q = 1 - p$. Then, for $\bar{P}_l(r)$, we have typically s success runs of length $\leq r - 1$, separated by $s - 1$ failures, where $l = s - 1 + \sum_{j=1}^s l_j$,

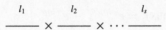

a probability of $q^{s-1} p^{\sum_1^s l_j}$. To sum up these probabilities at given l and r, we construct the generating function

$$\sum \bar{P}_l(r) z^l = \sum_s \sum_{\{l_j \leq r-1\}} q^{s-1} p^{\sum_1^s l_j} z^{s-1+\sum_1^s l_j}$$

$$= \sum_s (qz)^{s-1} \left[\sum_{l=0}^{r-1} (pz)^l \right]^s$$

$$= \sum_1^\infty (qz)^{s-1} \left[\frac{1 - (pz)^r}{1 - pz} \right]^s$$

$$= \left[\frac{1 - (pz)^r}{1 - pz} \right] \bigg/ \left\{ 1 - \frac{qz[1 - (pz)^r]}{1 - pz} \right\}$$

$$= [1 - (pz)^r]/[1 - z + qz(pz)^r].$$

This is a ratio of polynomials, $Q_{r-1}(z)/Q_r(z)$ (the factor $1 - pz$ cancels out and the subscript indicates order) and hence can be partial-fractioned as

$$\sum_{\alpha=1}^r a_\alpha/(z - z_\alpha)$$

where z_α ($\neq 1/p$) satisfies $1 - z_\alpha + q z_\alpha (p z_\alpha)^r = 0$ and

$$a_\alpha = \lim_{z \to z_\alpha} Q_{r-1}(z)(z - z_\alpha)/Q_r(z) = Q_{r-1}(z_\alpha)/Q'_r(z_\alpha)$$

$$= \frac{1 - (p z_\alpha)^r}{-1 + qr\,(p z_\alpha)^r} = \frac{z_\alpha - \frac{z_\alpha - 1}{q}}{-z_\alpha + r(z_\alpha - 1)}.$$

Now $\sum_{\alpha=1}^r a_\alpha/(z - z_\alpha) = \sum_{\alpha=1}^r \sum_l (-a_l/z_\alpha)\,(z^l/z_\alpha^l)$, and we conclude that

$$\bar{P}_l(r) = \sum_\alpha \frac{-a_\alpha}{z_\alpha^{l+1}}.$$

For very large l, only the z_α, called z^*, of smallest absolute value will contribute, so

$$\bar{P}_l(r) = \frac{z^* - \frac{z^*-1}{q}}{z^* - r(z^* - 1)}\, \frac{1}{z^{*l+1}}.$$

For z^*, assuming sufficiently large r, we simply iterate: $z^* = 1 + q z^*$ $(p z^*)^r = 1 + q p^r + \cdots$. Hence, to leading order, $\bar{P}_l(r) = \exp(-lq p^r)$, or

$$P_l(r = \max) = e^{-lq p^{r+1}} - e^{-lq p^r}$$
$$= e^{-lq p^r}\left(e^{lq^2 p^r} - 1\right)$$
$$= \exp\left[-lq p^r + \ln\left(e^{lq^2 p^r} - 1\right)\right].$$

This has a sharp maximum with respect to r, located at

$$-lq(\ln p)p^r + lq^2(\ln p)p^r\, e^{lq^2 p^r}/\left(e^{lq^2 p^r} - 1\right)$$
$$= lq(\ln p)p^r[-1 + q/(1 - e^{-lq^2 p^r})] = 0,$$

or $(1/p)^r = lq^2/(\ln 1/p)$. The derivative vanishes at the maximizing value,

$$\bar{r} = \ln(lq^2/\ln 1/p)/(\ln 1/p),$$

and for the second derivative we have

$$\frac{\partial}{\partial r}[(\ln p)lq]p^r\left(-1 + q/1 - e^{-lq^2 p^r}\right)$$
$$= lq(\ln p)p^r\left[-\ln p + q\ln p/1 - e^{-lq^2 p^r}\right.$$
$$\left. - lq^3(\ln p)e^{-lq^2 p^r}/\left(1 - e^{-lq^2 p^r}\right)^2 p^r\right]$$
$$= -l^2 q^4 (\ln p)^2\, p^{2r} e^{-lq^2 p^r}/q^2 = -l^2 q^2 (\ln p)^2\, p^{2r+1}$$
$$= -p/q^2 (\ln p)^4.$$

Expanding the exponent of $P_l(r = \text{max})$ yields

$$P_l(r = \text{max}) = C \exp -\frac{1}{2} \frac{p}{q^2} \left(\ln \frac{1}{p} \right)^4 (r - \bar{r})^2 ,$$

where C is a normalization constant, and we end up with

$$E(r_{\text{max}}) = 2 \frac{\ln n}{\ln 1/p} + \frac{\ln \left(\frac{1}{2} q^2 / \ln 1/p \right)}{\ln 1/p} + \cdots ,$$

$$\sigma^2(r_{\text{max}}) = (q^2/p)/(\ln 1/p)^4 .$$

See also Karlin and Ost (1988) and Mott et al. (1990). A brief derivation in a more general context will be given in Section 4.1.4.

As an immediate application, we can look at the full DNA (chosen as extremes) of SV40 and λ phage:

	$E(r_{\text{max}})$	r_{max} observed	σ
SV 40	12	72	2
λ phage	15	15	2

The 72-bp repeat is obviously significant.

Assignment 4

1. Suppose we collapse the DNA information into purines (0) and pyrimidines (1). Show that the χ^2 test for independence has $\nu = 1$ degrees of freedom, evidenced by the fact that all four of the observed $|N_\alpha^{1/2} y_\alpha|$ are the same.
2. How would you generalize the longest repeat result to accord with the order-1 Markov chain as random default?

3.2.2. Displaced Correlations

Let us examine the long-range structure more systematically. To get a feeling for the quantities of interest, suppose first that we are not far from the independent placement of bases (with the base type now denoted by $s, t, \ldots$). The most general question we might ask would be about the nature of the L-subsequence distribution, which, in view of the approximate independence, we write as

$$E[\delta_{x+1}(s_1)\,\delta_{x+2}(s_2)\cdots s_{x+L}(s_L)]$$

$$= Av_x \prod_{i=1}^{L} \left\{ p_{s_i} + \left[\delta_{x+i}(s_i) - p_{s_i} \right] \right\}$$

$$= \left(\prod_{i=1}^{L} p_{s_i} \right) Av_x \left\{ 1 + \sum_{i=1}^{L} \frac{\delta_{x+i}(s_i) - p_{s_i}}{p_{s_i}} \right.$$

$$\left. + \sum_{i<j\leq L} \frac{\left[\delta_{x+i}(s_i) - p_{s_i} \right]\left[\delta_{x+j}(s_j) - p_{s_j} \right]}{p_{s_i} p_{s_j}} + \cdots \right\}$$

$$= \left(\prod_{i=1}^{L} p_{s_i} \right) \left[1 + \sum_{i<j\leq L} \frac{p_{s_i s_j}\,(j-i) - p_{s_i} p_{s_j}}{p_{s_i}\,p_{s_j}} + \cdots \right].$$

Here we have taken advantage of the definition $p_s = Av_x\,\delta_x(s)$, as well as of the translation invariance – to within end effects – of the averaging, which allows us to write

$$Av_x[\delta_{x+i}(s)\,\delta_{x+j}(t)] \equiv p_{st}\,(j-i) \quad \text{for } i < j.$$

Thus it is sufficient at this level of analysis to know the *covariances*,

$$C_{ss'}(n) = \text{Cov}[\delta_x(s),\ \delta_{x+n}(s')]$$
$$= p_{ss'}(n) - p_s p_{s'},$$

the mean product of base type s, and type s', fluctuations when n sites apart. In fact, we now phrase a set of significant questions directly in terms of the $C_{ss'}(n)$, without approximation.

3.2.3. Nucleotide-Level Criteria

Imagine then a long fragment of a single strand of DNA. We expect that for the whole fragment, p_A, p_C, p_G, p_T will be different, although it is true that if we average over fragments of two-strand complementary DNA, we should find that $p_A = p_T$ and $p_G = p_C$ In fact, these equalities tend to hold even for a long-enough piece of a single strand, the so-called *strand-symmetry* condition, e.g., consider (Zhang and Marr, 1994b) the completely known 315-kb single strand of the yeast *Saccaromyces cerevisiae* Chromosome III (predominantly exon), in which we find $p_A = 0.31$, $p_T = 0.30$, $p_G = 0.19$, and $p_C = 0.20$. We might attribute this symmetry, for example, to anomalous crossover events that result in frequent inclusion of segment inversions.

In greater detail, we want to study base correlations, the covariances $C_{st}(n)$, or the corresponding correlation coefficients

$$r_{st}(n) = p_{st}(n) / p_s p_t.$$

The first question is that of persistence, the way that $r_{ss}(n)$ decays on the

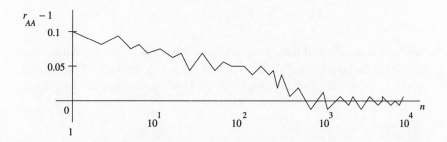

average to its random value of 1. All four bases in this chromosome behave similarly in this respect, with a range of $\sim$1500 nt (nucleotides). r_{AA} and r_{TT} are very similar, whereas r_{CC}, similar to r_{GG}, ends with higher amplitude fluctuation. It is not possible at this level to distinguish among detailed decay forms for $r_{ss} - 1$, such as $e^{-\alpha n}$ and $n^{-\alpha}$.

Going on to displaced cross correlations $r_{st}(n)$, it is useful to distinguish

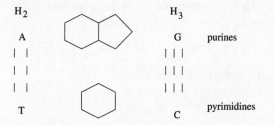

on the basis of pairing type, two or three hydrogen bonds, and heterocyclic type, purine or pyrimidine. A first result is that $r_{pu,pu'}(n)$ and $r_{py,py'}(n)$ have negative correlations ($r - 1 < 0$) for $n > 4$, with a range of $\sim$150 nt. For $r_{py,py'}(n)$ and $r_{py',pu}(n)$, with pairs not complementary in the sense of pairing type either (i.e., A and C, G and T), we start with small-n negative correlations, which then reverse. Finally, for a complementary pair, there is small negative correlation, up to $\sim$150 nt. Although the wide fluctuations do not let us say much more, there is one significant regularity in the data: On the average,

$$r_{st}(n) = r_{\bar{t}\bar{s}}(n)$$

(where the overbar indicates complementary base), a much more detailed
strand symmetry that would be exactly true for a pair of complementary
strands.

We obtain some reduction in fluctuations by further amalgamating the
bases. One such amalgamation or figure of merit is the mutual information
function of Li (1990),

$$M(n) = \sum_{s,t} p_{st}(n) \ln_2 [p_{st}(n)/p_s \, p_t]$$

which vanishes if and only if $p_{st}(n)$ factorizes as $p_s \, p_t$. In the human co-
agulation factor (HUMCF) VII of 12,830 nt ($\sim$0.76 intron, 0.11 exon), this
gives an extended correlation range of $\sim$800 nt, consistent with the general

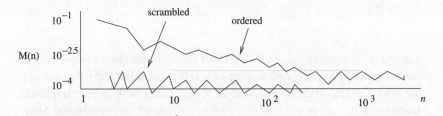

observation that introns tend to have long correlations. Compared with the
scrambled sequence of the same base ratios, but randomly placed bases, the
result is quite sharp. Another statistical quantity used (Voss, 1992) that is
symmetric over the bases

$$C(n) = \text{tr} \, C_{st}(n) = \sum \left[p_{ss}(n) - p_s^2 \right]$$

has quite similar behavior, as does (Li et al., 1994)

$$\bar{C}(n) = \sum_{s,t} C_{st}(n) \, u_s \cdot u_t,$$

where the vector u_t is one of the four vertices of a tetrahedron at distance 1
from the origin (clearly $u_s \cdot u_s = 1$, $u_s \cdot u_t = -\frac{1}{3}$ for $s \neq t$).

Expansion–Mutation Model. There is no particular reason for total symmetry, and so we may choose to only distinguish between dual pairs,

$$C_d(n) = \frac{1}{2}[C_{A+T,A+T}(n) + C_{G+C,G+C}(n)]$$

(where $C_{s+t,s+t} = C_{ss} + C_{st} + C_{ts} + C_{tt}$), or heterocyclic type

$$C_h(n) = \frac{1}{2}[C_{A+G,A+G}(n) + C_{C+T,C+T}(n)].$$

Of course, because $\delta_x(A) + \delta_x(G) = 1 - \delta_x(C) - \delta_x(T)$ at each site, it is clear that $C_{A+G,A+G}(n) = C_{C+T,C+T}(n)$. It is similarly clear that, in terms of the variance, e.g.,

$$V_{C+T-A-G}(n) = C_{C+T-A-G,C+T-A-G}(n),$$

we have

$$C_h(n) = \frac{1}{4} \, V_{C+T-A-G}(n) = V_{A+G}(n).$$

In either event, we are reduced to a sequence of two symbols, say $\delta_x(A) + \delta_x(G) = 0$ or 1, and ask for the genesis of extended persistence. A number of models have been suggested, in particular for what is claimed (Peng et al., 1992) to be a fractal or power-law decay $n^{-\alpha}$ dominated by noncoding DNA. One of these (Li and Kaneko, 1992) is evolutionary expansion with error. In its most primitive version, this is a simple cellular automaton (Lindenmayer, 1968) model with stochastic evolution: For a string of 0s and 1s at some point in time, a normal step is a doubling of an element, $0 \to 00$ or $1 \to 11$, but a certain fraction of the time (p) this is replaced with an error, $0 \to 1$ or $1 \to 0$. Thus there is a mean expansion rate per site of $k = 2(1 - p) + p = 2 - p$ at each time step. A heuristic consequence is this. Let $p_{st}^{(N)}(n)$ be the probability that a pair st is separated by n after N time steps. Then assume some transition law (now s and t are 0 or 1)

$$p_{st}^{(N+1)}(n) = \sum_{s't'n'} T_{stn,s't'n'} \, p_{s't'}^{(N)}(n').$$

For $N \to \infty$, we will have $p_s \, p_t = \sum T_{stn,s't'n'} p_{s'} \, p_{t'}$, so that the covariance $C_{st}^{(N)}(n) = p_{st}^{(N)}(n) - p_s \, p_t$ satisfies the same equation. However, the transition from n' to n will be dominated by the same expansion rate as the system: $n \sim kn'$; thus we can write instead (assuming no further n dependence)

$$C_{st}^{(N+1)}(n) = \sum_{s't'} T_{st,s't'} \, C_{s't'}^{(N)}(n/k).$$

If $\lambda > 1$ is the maximum eigenvalue of the 2×2 matrix T and v is the corresponding eigenvector, then asymptotically we must have

$$\bar{C}^{(N+1)}(n) = \lambda \, \bar{C}^{(N)}(n/k),$$

where $\bar{C}$ is the v component of C. This has the obvious stationary solution

$$\bar{C}(n) \propto n^{\ln \lambda / \ln k},$$

the desired power law. This may indeed be valid in prebiotic expansion and in larger entities that are repeated, with mistakes.

Simple Sequence Repeats. In the preceding subsection, the symbols 0, 1, and more might refer to small subsequences that are duplicated. Hence we might not be surprised to find many repeats, such as $(GT)^n$ with n as large as 20, although their probable number in the two-strand human genome under random equivalent placement of bases would be only $\sim 2 \times (3 \times 10^9)/4^{20} \sim 0.005$. Actually, such microsatellites with $n > 10$ occur $\sim 10^5$ times in the human genome; they are highly polymorphic in length and provide a nongenetic signature of an individual much used for forensic purposes. What sort of length distribution would we expect?

Suppose (Bell, 1994) that a seed of n_0 repeats of length l, with $m = n_0 l$, is required for viability and that this appears by mutation from a "neighboring" sequence. The latter must be correct at $m - 1$ positions and wrong at 1 of the m positions, and so will occur with probability $p_1 = (3/4) m (1/4)^{m-1}$. In the counting, a sequence $(TG)^n$ will be equivalent to $(GT)^n$, and also (on the other strand) to $(CA)^n$ or $(AC)^n$ – an equivalence class of $N_{eq} = 4$ [whereas $(AT)^n$ would have $N_{eq} = 2$]. Also, a next-upstream entry of the last base of a repeat would signal a different repeat and not be counted (probability $3/4$ that the desired repeat is recognized), whereas a downstream repeat of the repeat unit [probability $(1/4)^l$] would be entered as a new value of n. Hence the actual probability to be used is

$$p_1' = \left(\frac{3}{4}\right)^2 m \, N_{eq} \left(\frac{1}{4}\right)^{m-1} \left[1 - \left(\frac{1}{4}\right)^l\right].$$

For G base pairs, there are G possible starting positions for the sequence (ignoring end effects), so if R is the mean base substitution rate – hence $R/3$ for a correct substitution – then we have a source of n_0-fold repeats given by

$$S = \frac{1}{3} \, G p_1'.$$

Thus, for $(GT)^{n_0}$, knowing that $G = 3 \times 10^9$ bp and $R \sim 5 \times 10^9$ per bp/year, we would have $S = 0.66$ (genome/year) for $n_0 = 2$, 0.062 for $n_0 = 3$, etc.

Given the source of n_0-fold repeats, we need a model for the dynamics. With $n = n_0 + k - 1$, so that $k = 1$ initially, imagine a birth–death process of λ/unit time per excess repeat, i.e., probability λk/unit time for $k \rightarrow k + 1$ and also λk for $k \rightarrow k - 1$. Thus, if $P(1, k, t)$ is the probability of having n repeats at time t, given that we started with $k = 1$, counting the disappearance from the k pool, the arrivals from $k - 1$, and the arrivals from $k + 1$, we clearly have

$$\frac{d}{dt} P(1, k, t) = - 2\lambda k\, P(1, k, t) + \lambda(k - 1)\, P(1, k - 1, t)$$
$$+ \lambda(k + 1)\, P(1, k + 1, t)$$

for $k = 1, 2, \ldots$. On the other hand, if k descends to 0, the repeat is not feasible and the process stops,

$$\frac{d}{dt} P(1, 0, t) = \lambda\, P(1, 1, t),$$

and of course we have the initial condition

$$P(1, k, 0) = \delta_{k,1}.$$

We solve the difference equation in standard fashion by setting up the generating function

$$P(z, t) = \sum_{k=0}^{\infty} P(1, k, t)\, z^k,$$

multiplying the difference equation by z^k, summing over $k = 1, 2, \ldots$ and adding its $k = 0$ partner; we find

$$\frac{\partial}{\partial t} P(z, t) = \lambda(z - 1)^2\, P(z, t),$$

$$P(z, 0) = z.$$

This is readily solved as

$$P(z, t) = 1 + \frac{z - 1}{1 - (z - 1)\lambda t},$$

yielding the distribution

$$P(1, k, t) = \frac{(\lambda t)^{k-1}}{(1 + \lambda t)^{k+1}} \quad \text{for} \quad k \geq 1$$

[so that $P(1, 0, t) = \lambda t / (1 + \lambda t)$]. With a steady source S of seeds from $t = 0$ to $t = T$, the total number at T is

$$N(1, k, T) = S \int_0^T P(1, k, t)\, dt = \frac{S}{\lambda k} \left(\frac{\lambda T}{1 + \lambda T} \right)^k,$$

the full distribution. Note that the total number of repeat sequences is

$$N(T) = \sum_{k=1}^{\infty} N(1, k, T) \sim \frac{S}{\lambda} \int_1^{\infty} \frac{1}{k} \, e^{-k/\lambda T} \, dk$$

$$= \frac{S}{k} \, E_1 \left(\frac{1}{\lambda T} \right),$$

in terms of the exponential integral E_1 and has the asymptotic large-T form

$$N(T) \sim \frac{S}{\lambda} (\ln \lambda T + \gamma)$$

($\gamma = 0.577 \ldots$ is Euler's constant), a very slow increase even in the absence of mutational deterioration.

Length Distributions. Repeats need not be so simple, need not be in tandem, and need not be confined to noncoding regions. For example, it has been suggested on numerous occasions that not only are genes just a shuffling of a smaller number of exons (Gilbert, 1997; Stoltzfus et al., 1994) (but see Doritt et al., 1990), but also (see, e.g., Dwyer, 1998) that exons are composed of relatively small numbers of ancestral units, of course evolved to some extent. One thing is certain: The three-letter codons are highly repeated, but are not equally likely, and this can very much affect the correlation structure within an exon. That is, if codon (u, v, w) occurs at relative frequency f_{uvw}, then, neglecting any correlations between codons in a very long exon, we will have (Herzl and Grosse, 1997)

$$p_{st}^{\infty}(3k) = \frac{1}{3} \sum_{\substack{vw \\ v'w'}} (f_{svw} \, f_{tv'w'} + f_{vsw} \, f_{v'tw'} + f_{vws} \, f_{v'w't})$$

independently of k, and

$$p_{st}^{\infty}(3k+1) = \frac{1}{3} \sum_{\substack{uv \\ u'w'}} (f_{svw} \, f_{u'tw} + f_{usw} \, f_{u'w't} + f_{uws} \, f_{tu'w'}) = p_{ts}^{\infty}(3k-1).$$

Thus a symmetric probe such as the mutual information function $M(n)$ will simply have a fine structure with period 3.

For a distribution of finite exon lengths $p(l)$, matters are more interesting. We first note that for weak dependence, i.e., small $C_{st}(n) = p_{st}(n) - p_s \, p_t$, the mutual information can be expanded as

$$M(n) = \frac{1}{2} \sum_{s,t} C_{st}(n)^2 / p_s \, p_t.$$

Now suppose, as in prokaryotes, we have almost all exons. Then the dominant correlations are still for a pair within an exon, but there are only $l - n$ possible placements, compared with $L - n \sim L$ for the full-length L sequence, so that

$$C_{st}(n) = \sum_{l=n}^{\infty} \rho(l)(l - n)/L \; C_{st}^{\infty}(n).$$

The factor $z(n) = \sum \rho(l)(l - n)/L$ produces an exponential distribution if $\rho(l)$ is exponential and a power law $\propto n^{2-\beta}$ if $\rho(l) \propto^{-\beta}$ above some cut-off, and so in fact the long-range correlation represented by $\mu(n)$ is a direct reflection of the exon length distribution.

The length distribution of ORF's (open reading frame = DNA stretch between a start codon and a stop codon in the same frame = coding sequence) in a number of organisms has been studied (Li, 1999), with the conclusion that these are always exponential, although the exponent may change abruptly at a few $\times 10^2$ bp. Thus the exon contribution to asymptotic correlations will only be exponential. On the other hand, it was noted (Almirantis and Provata, 1999) that there is empirical evidence that the length distribution of purine and pyrimidine clusters has the large-l form

$$\bar{\rho}(l) \propto l^{-1-u},$$

which suggests that the distribution be modeled as a stable distribution (Feller, 1950). If this is the case, then a noncoding region, modeled as a concatenation of Pu and Py clusters, will have the same asymptotic length distribution, with the same consequences for sets of introns.

3.2.4. Batch-Level Criteria

To improve the statistics, we should use bigger batches of data. Most obvious is to accumulate information in *windows* of bases, $W_N(x)$, N bases starting at x, and average over x (Fickett and Tung, 1992). The windows, as expanded points, should be made much shorter than the fragment being examined and will then give meaningfully resolved statistics. We distinguish between correlations *within* a window and between windows.

Within-Window Correlations

For a window of size N, we tally #s, the number of occurrences of s in a given window, and construct

$$\mathrm{Cov}^{(N)}(\#s, \#t) = \mathrm{Av}_x(\#s - \langle \#s \rangle)(\#t - \langle \#t \rangle).$$

In practice, the xs are usually chosen so that the windows do not overlap, thereby avoiding overt correlations between windows, but it turns out empirically that none of the significant results alter if we let x run over all sites, and we shall do so. Of course, $\langle \#s \rangle = N p_s$, but also, because $\#s = \sum_{r=0}^{N-1} \delta_{x+r}(s)$, we have $\text{Cov}^{(N)}(\#s, \#t) = \sum_{r=0}^{N-1} \sum_{r'=0}^{N-1} \text{Cov}[\delta_{x+r}(s), \delta_{x+r'}(t)]$, or, gathering together common values of $|r - r'| = n$,

$$\text{Cov}^{(N)}(\#s, \#t) = N(p_s \delta_{s,t} - p_s p_t) + \sum_{n=1}^{N-1} (N - n)[C_{st}(n) + C_{ts}(n)] \,,$$

and, as an important special case, the variance

$$\text{Var}^{(N)}(\#s) = \text{Cov}^{(N)}(\#s, \#s) = N p_s(1 - p_s) + 2 \sum_{1}^{N-1} (N - n) C_{ss}(n) \,.$$

A convenient construct is the *correlation coefficient* of standard statistical usage (not to be confused with the correlation coefficient $r_{st}(n)$ of Subsection 3.2.3)

$$\text{corr}^{(N)}(\#s, \#t) = \text{Cov}^{(N)}(\#s, \#t) / [\text{Var}^{(N)}(\#s) \, \text{Var}^{(N)}(\#t)]^{1/2} \,,$$

where it is readily verified that $-1 \leq \text{corr}^{(N)} \leq 1$. Note that if the base placement were independent (Bernoulli), so that $C_{st}(n) = 0$, we could have

$$\text{Var}^{(N)}(\#s) = N p_s(1 - p_s),$$

$$\text{corr}^{(N)}(\#s, \#t) = -[p_s/(1 - p_s)]^{1/2} [p_t/(1 - p_t)]^{1/2} \quad \text{for } s \neq t$$

(the latter being $-1/3$ when all $p_s = 1/4$).

Let us look a bit at $\text{Var}^{(N)}(\#s)$ [similar considerations apply to $\text{Cov}^{(N)}(\#s, \#t)$]. If the correlation is short range, e.g., $C_{ss}(n) \sim A e^{-\alpha n}$, as in Markov order 1, then $\sum (N - n) C_{ss}(n) \sim \int_0^N (N - n) A e^{-\alpha n} dn = A e^{-\alpha N}$ $(\partial/\partial\alpha) \int_0^N e^{\alpha n} dn = (A/\alpha)N - A[(1 - e^{-\alpha N}/\alpha^2]$, so that asymptotically

$$\text{Var}^{(N)}(\#s) = N \left[p_s(1 - p_s) + 2 \frac{A}{\alpha} \right] + \cdots$$

still varies as N. On the other hand, if the decay is slower, e.g., the power law $C_{ss}(n) \sim A n^{-\alpha}$, where $0 < \alpha < 1$, then $\int_0^N (N - n) A n^{-\alpha} dn = A/(1 - \alpha)$ $(2 - \alpha) N^{2-\alpha}$, and

$$\text{Var}^{(N)}(\#s) \sim \frac{2A}{(1 - \alpha)(2 - \alpha)} N^{2-\alpha} + p_s(1 - p_s) N + \cdots$$

is dominated by a power higher than N^1.

What happens in animal genomes? Typically the extremes of human (usually with the c-DNA which is complementary to the processed RNA that codes for proteins, excluded as being special) and *E. coli* bacteria are studied. The first observation is that variances are much larger than those of Bernoulli or even any reasonable (e.g., 10) Markov order. For example, at $N \sim 1000$, we have Var $\sim 10,000$ for the full human combined $A + T$ content, 2300 for

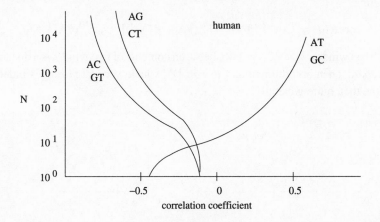

corresponding *E. coli*, in a kilobase-pair window, compared with ~ 250 for independence. There are also dramatic effects in correlations (Fickett and Tung, 1992) that start at $\sim -1/3$, as expected for small windows, but quickly depart to extreme values, and by 1.2 kb have arrived at

$$\text{corr(A, C)} = -0.78, \ \text{corr(A, G)} = -0.65, \ \text{corr(A, T)} = 0.42,$$
$$\text{corr(G, T)} = -0.77, \ \text{corr(C, T)} = -0.67, \ \text{corr(C, G)} = 0.45.$$

The AT and GC correlations quickly becomes larger and positive, perhaps approaching the corr(A, T) $= 1 =$ corr(G, C) for full strand symmetry in which #A $=$ #T and #G $=$ #C. This justifies focusing on A $+$ T or G $+$ C content, ironing out fluctuations, without losing information.

There is as well the qualitative fact that closely

$$\text{corr}(\#s, \#t) = \text{corr}(\#u, \#v),$$

where $s, t, u,$ and v cover all four base types, in any order – an equality that holds to $\sim 1\%$ when we look at the corresponding covariances. This is actually not too informative. At fixed N, $\#s + \#t = -(\#u + \#v) + N$, so Var $(\#s + \#t) = \text{Var}(\#u + \#v)$ or $\text{Var}(\#s) + 2 \text{Cov}(\#s, \#t) + \text{Var}(\#t) = \text{Var}(\#u) + 2\text{Cov}(\#u, \#v) + \text{Var}(\#v)$, equivalent to

$$\text{Var}(\#A) = \text{Var}(\#C) = \text{Var}(\#G) = \text{Var}(\#T),$$

which is hardly surprising.

Between-Window Correlations

Here the persistence effect noted in the previous subsection again appears. Define

$$\text{corr}_{AT}(d) = \text{Cov}^{(N)}[\#A + \#T, \ (\#A + \#T)(d)] \, / \, \text{Var}^{(N)}(\#A + \#T)$$

at fixed window size N, say 1 kb, between contents of two windows a distance d apart. Then both human and *E. coli* DNA have very large ranges indeed. Note that, quite generally,

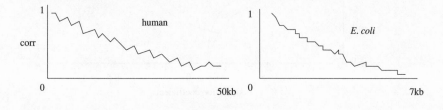

$$\text{Cov}^{(N)}[\#s, \#t(d)] = \sum_{r,r'=0}^{N-1} \text{Cov}[\delta_{x+r}(s), \delta_{x+r'+d}(t)]$$

$$= \sum_{-N}^{0}(n + N)C_{st}(d + n) + \sum_{1}^{N}(n - N)C_{st}(d + n)$$

$$= N^2 \bar{C}_{st}(d) \, ,$$

where $\bar{C}_{st}(d)$ is a triangularly weighted average of covariances at separations from $d - N$ to $d + N$ (of course, $d > N$).

Factorial Moments

Higher within-window moments of $\#s$, or of combined bases, e.g., $\#s + \#t$, have been considered as well (Mohanty and Narayana Rao, 2000), and in particular the normalized factorial moments

$$F_q^{(N)}(\#s) = \langle (\#s)!/(\#s - q)! \rangle \, / \langle \#s \rangle^q$$

have some desirable properties. For example, if $\#s$ is Poisson distributed, $\text{Pr}(\#s) = \lambda^{\#s}/(\#s)! \, e^{-\lambda}$, then all $F_q^{(N)}(s) = 1$. For a number of test DNA sequences, both with and without introns, there appears to be a monotonic trend as a function of the window size N: for $N < N_c \sim 10^2 - 10^3$ bp, all $F_q^{(N)} < 1$, variances increase as N and fluctuations appear Gaussian; for $N > N_c$, $F_q^{(N)} > 1$, variances increase as $N^{1.4-1.5}$ and fluctuations appear to be non-Gaussian. The Gaussian property can be assessed (Allegrini et al., 1998) by means of the kurtosis

$$\eta(N) = 1 - 3\frac{\langle U(N)^2\rangle_x^2}{\langle U(N)^4\rangle_x},$$

where $U(N)$ is the window-N sum $U(N) = \sum_{r=1}^{N} u(x+r)$ of $u(x) = \{\pm 1$ as the base at x is a pyrimidine or purine}. For a Gaussian distribution, $\eta(N) = 0$, whereas a model of correlated base sequences embedded in a sea of uncorrelated bases reproduces observations of the preceding type quite well.

3.2.5. Statistical Models

Patch Model

We return to the question of the genesis of long-range correlations, as assessed again by means of the window-N quantities

$$\#u = \#C + \#T - \#A - \#G,$$

whose per-site value is $u = +1$ for a pyrimidine and $u = -1$ for a purine (Karlin and Brendel, 1993). As we have seen, $\mathrm{Var}_u(n) = 4\,\mathrm{Var}_{CT}(n)$, and so $\mathrm{Var}^{(N)}(\#u) = 4\,\mathrm{Var}^{(N)}(\#C + \#T)$ as well, but let us stick to the variable $\#u$ instead, which, as N increases, executes a simple random walk, $\Delta(\#u) = \pm 1$ at each step, on the integers. We have, in the usual way (but noting that here $u^2 = 1$),

$$\mathrm{Var}^{(N)}(\#u) = N + 2\sum_{1}^{N-1}(N-n)\,C_{uu}(n).$$

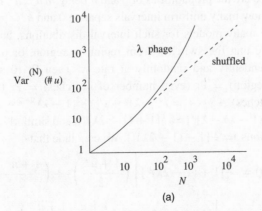

(a)

A plot of the bacteriophage λ sequence shows an increase of faster than N^1 compared with the shuffled sequence, which indeed goes as N. In fact, a

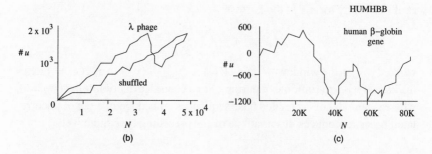

(b)   (c)

simple count of $\#u(N)$ at a fixed starting position shows an expected linear
increase for shuffled λ phage compared with piecewise linear sections for
λ itself and even more pieces for a section of human genome, looking as
if we have patches of different uniform u (or CT) composition, i.e., small-
scale isochores. We can model this situation most simply by supposing two
types of patch or subregion, one with $\langle u \rangle = u^+$, the other with $\langle u \rangle = u^-$.
The probabilities that $u = +1$(CT) and that $u = -1$(AG) are, respectively,
$\frac{1}{2}(1 + u^\pm)$ and $\frac{1}{2}(1 - u^\pm)$ in these two regions. Now suppose that there are
no correlations *within* each region, i.e., Bernoulli or Markov order 0. Then
the joint probabilities for $u = 1$ or -1 at sites 0 and n, conditional on 0 and
n being in $u^\pm$ regions (the first and the second arguments $u^\pm$ are distinct) are
given by

$$
\begin{bmatrix}
p_{11}(n \mid u^\pm, u^\pm) & p_{1-1}(n \mid u^\pm, u^\pm) \\
p_{-11}(n \mid u^\pm, u^\pm) & p_{-1-1}(n \mid u^\pm, u^\pm)
\end{bmatrix}
=
\begin{bmatrix}
\frac{1+u^\pm}{2} \frac{1+u^\pm}{2} & \frac{1+u^\pm}{2} \frac{1-u^\pm}{2} \\
\frac{1-u^\pm}{2} \frac{1+u^\pm}{2} & \frac{1-u^\pm}{2} \frac{1-u^\pm}{2}
\end{bmatrix}.
$$

What we need are the probabilities of 0 and n being in $u^\pm$, $u^\pm$ regions, which
depends on how many uniform intervals separate 0 and n.

There are many models for such interval distributions, but the simplest
is to suppose that the switches from region to region, or patch to patch,
occur independently and randomly at rate λ. Then Pr (0 and n both in
same type region) = Pr (even number of switches) = Pr (no switch) +
Pr (two switches) $+ \cdots + = (1 - \lambda)^n + \binom{n}{2}\lambda^2(1 - \lambda)^{N-2} + \cdots = \frac{1}{2}\{[(1 - \lambda) + \lambda]^n + [(1 - \lambda) - \lambda]^n\} = \frac{1}{2}[1 + (1 - 2\lambda)^n]$, and similarly Pr (0 and n on
different regions)$= \frac{1}{2}[1 - (1 - 2\lambda)^n]$. We conclude that

$$
\begin{aligned}
p_{11}(n) &= \frac{1}{4}[1 + (1 - 2\lambda)^n]\left[\left(\frac{1+u^+}{2}\right)^2 + \left(\frac{1+u^-}{2}\right)^2\right] \\
&\quad + \frac{1}{4}[1 - (1 - 2\lambda)^n]\left(2\frac{1+u^+}{2}\frac{1+u^-}{2}\right) \\
&= \frac{1}{4}\left(1 + \frac{u^+ + u^-}{2}\right)^2 + \frac{1}{4}\left(\frac{u^+ - u^-}{2}\right)^2(1 - 2\lambda)^n,
\end{aligned}
$$

so that $C_{11}(n) = p_{11}(n) - p_{11}(\infty) = \frac{1}{16}(u^+ - u^-)^2(1 - 2\lambda)^n$. It follows that

$$\text{Var}^{(N)}(\#u) = N + \frac{1}{8}(u^+ - u^-)^2 \sum_1^{N-1}(N - n)(1 - 2\lambda)^n .$$

In fact, for large patches, $\lambda \sim 0$, we have

$$\text{Var}^{(N)}(\#u) = N + \frac{1}{16}(u^+ - u^-)^2 N(N - 1),$$

which, in form, matches the observed N dependence as well as the "fractal" $N + \gamma N^{2-\alpha}$.

However, quantitative comparison is not reliable, and evidence has accumulated that the long-range power-law dependence of correlations is indeed the norm. In particular, Peng et al. (1994) (see also Bernaola-Galván et al., 1996), have examined in greater detail the extent to which a patch model – and there certainly are patches – can mimic long-range power-law correlations. They did this by first dividing the sequence into subsequences of length $l \sim 10^2$, small compared with anticipated patches, finding an optimal linear fit to $\#u(N)$ in each subregion, and subtracting it out (detrending), thus eliminating any patch bias. The resulting $\#u_{\text{det}}(N)$ hence fluctuates around zero, and it is to this that the window standard-variation analysis is applied. When this is done, the power-law variation is recovered as well as the threshold l that measures the entree of patch contributions.

With confidence that the power-law variation is intrinsic, the analysis was formalized (Lu et al., 1998) in terms of the Hurst index for self-similar patterns. That is, if $X^{(m)}$ is a measured characteristic averaged over a block of length m, and the system is divided into concatenated blocks indexed by i, then the assertion that

$$r^{(m)}(k) = \langle [X_i^{(m)} - \bar{X}][X_{i+k}^{(m)} - \bar{X}] \rangle_i = k^{-\beta} \Lambda(k)$$

with slowly varying Λ, independently of m for large m, defines a Hurst index $H = 1 - \frac{1}{2}\beta$ for the self-similar sequence in question. Actually, a related assessment, that of the dimensionality of a walk constructed with A, T, G, and C as ± 1 steps on coordinate axes in the plane, had been carried out much earlier (Berthelsen et al., 1992). The conclusion (implicitly by use of the approximate equality of A and T steps and G and C steps) was that dimensionality could be defined and is anomalously small compared with that of a random sequence, consistent with power-law asymptotic correlations.

Hidden Markov Models

The Karlin–Brendel model is a special case of a class of stochastic models (Gusfield, 1997; Baldi and Brunak, 1998; Durbin et al., 1998) that is being used more and more to represent partially random data economically, e.g., in pattern recognition. We imagine a "hidden" set of main states $\{s_i \in S\}$ and associated first-order Markov transition probabilities $\{T_{ss'} = P_r(s \mid s')\}$. The process can be regarded either as starting with a state distribution P_{s_1} or,

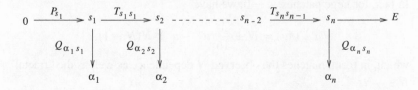

equivalently, as having a fixed starting state 0 appended to the set S of possible states, with $T_{s_1 s_0} = P_{s_1} \delta_{s_0,0}$. Likewise, we may stop it "by hand" at the nth stage or we can append an end state E and associated stopping probabilities T_{Es}, with the convention $T_{sE} = \delta_{s,E}$, thus producing a probabilistic stopping point.

To complete the description, there are the observed or "emitted" states $\{\alpha_i \in A\}$ that are independently chosen by means of the transition probabilities $Q_{\alpha s} = P_r(\alpha \mid s)$. Hence, at fixed n,

$$P_r(\alpha_1, \ldots \alpha_n) = \sum_{s_1, \ldots s_n} Q_{\alpha_n s_n} T_{s_n s_{n-1}} Q_{\alpha_{n-1} s_{n-1}} \cdots$$

$$\ldots T_{s_2 s_1} Q_{\alpha_1 s_1} P_{s_1},$$

and we can as usual inquire about correlations such as

$$p_{ss'}(n) = \sum_{x=1}^{n-m} P_r(\alpha_x = s, \alpha_{x+m} = s').$$

There are two extreme classes of transition matrices, the first being that of recurrent matrices, in which $T_{ss'} \neq 0$ for all main states. For example, in the Karlin–Brendel model, with two isochore types,

$$T = \begin{bmatrix} 1 - \lambda & \lambda \\ \lambda & 1 - \lambda \end{bmatrix},$$

with start and stop included by extending this to

$$T = \begin{bmatrix} 0 & 0 & 0 & E \\ 0 & \gamma(1 - \lambda) & \gamma\lambda & 0 \\ 0 & \gamma\lambda & \gamma(1 - \lambda) & 0 \\ 0 & 1 - \gamma & 1 - \gamma & 0 \end{bmatrix}.$$

For the emission process, we would have

$$Q = \begin{bmatrix} \frac{1}{2}(1 - u^-) & \frac{1}{2}(1 - u^+) \\ \frac{1}{2}(1 + u^-) & \frac{1}{2}(1 + u^+) \end{bmatrix}.$$

A second class is termed left to right. Here, we are looking for an underlying pattern, say $s^1, s^2, \ldots, s^m$, and so we have main states $s^1, s^2, \ldots, s^m$, and X (for anything), with transitions restricted to $X \rightarrow X$ or s^1, but $s^1 \rightarrow s^2 \rightarrow \cdots \rightarrow s^m \rightarrow X$. A typical problem associated with a such a hidden Markov model (HMM) is that of extracting the most likely hidden state sequence $\{s_i\}$ given the output $\{\alpha_i\}$, which we may accomplish, e.g., by first maximizing $Pr(\alpha_1, \ldots \alpha_n \mid T, Q)$ to find optimal $\{T, Q\}$, and then maximizing $Pr(s_1, \ldots s_n; \alpha_1 \ldots \alpha_n)$.

HMMs are accessible examples of a broad class of models ((a) in the figure below) in which coupled outputs are controlled by hidden coupled inputs.

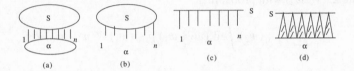

In (b) in the figure the outputs are coupled only to the inputs, i.e., Bernoulli, and we have a highly reduced model only acceptable at a sufficiently coarse level of resolution. The HMM, (c) in the figure, is a further specialization. At a finer level, the outputs should presumably at least be coupled informationally to their neighbors, (d) in the figure, and this is the thought behind the next model we consider.

Walking Markov Model

The hallmark of the Karlin–Brendel model was a set of "hidden" instructions, namely a subdivision into distinct homogeneous regions that was itself statistically determined. We therefore had a Bernoulli sampling with parameters directed by an independent interval process, which is in fact a special case of a Markov process (strictly speaking, "process" refers to a continuous iteration of the transition event, valid here only if the base-to-successive-base interval is regarded as infinitesimal). To do a better job on short-range structure, we would probably want each homogeneous subregion to be at least a first-order Markov chain ("chain" means a discrete iteration, and the interval between bases is certainly not small for low-order Markov). To do a better job on the long-range structure, we could replace the independent intervals or patches with a general first-order Markov process, making transitions between various types of patches (Elton, 1974) but, equivalent and more readily analyzed, we choose a model (Churchill, 1989; Fickett and Tung, 1992) in which a transition is attempted, with very low frequency, at every base, thereby succeeding on only a larger scale. The model of Fickett and Tung that we now turn to also focuses on $A + T$ occupation, consistent with the high degree of correlation between A and T.

Imagine then a first-order base-to-base Markov chain that depends on a hidden parameter w (or parameter set) that varies from step to step. The assumption is that w wanders autonomously by small increments, with transition probability $W(w \mid w')$. Then at each site there is a combination of base s and parameter w – a Markov chain on a higher space – and we can proceed sequentially, starting, say with (r, w_0), at site 0:

site 0: (r, w_0)
site 1: (s_1, w_1) with probability $p_{s_1|r}(w_1) \, W(w_1 \mid w_0)$
site 2: (s_2, w_2) with probability

$$\sum_{s_1} p_{s_2|s_1}(w_2) \int W(w_2 \mid w_1) \, p_{s_1|r}(w_1) \, W(w_1 \mid w_0) \, dw_1$$

$$\cdots$$
$$\cdots$$
$$\cdots$$

and at site N, the probability of (s, w), given (r, w_0) initially, is

$$p_{s|r}^{(N)}(w \mid w_0) = \sum_t p_{s|t}(w) \int W(w \mid w') \, p_{t|r}^{(N-1)}(w') \, dw'$$

where $p_{s|r}^{(0)}(w \mid w_0) = \delta_{rs} \, \delta(w - w_0)$.

Now we specialize to the two-state context, AT versus GC. A 2×2 Markov transition matrix must have the form $\begin{bmatrix} 1-\alpha & \beta \\ \alpha & 1-\beta \end{bmatrix}$, so that

$$\begin{bmatrix} p_{AT|AT}^{(N)}(w \mid w_0) \\ p_{GC|AT}^{(N)}(w \mid w_0) \end{bmatrix}$$

$$= \begin{bmatrix} 1 - \alpha(\omega) & \beta(w) \\ \alpha(w) & 1 - \beta(w) \end{bmatrix} \int W(w \mid w') \begin{bmatrix} p_{AT|AT}^{(N-1)}(w' \mid w_0) \\ p_{GC|AT}^{(N-1)}(\omega' \mid \omega_0) \end{bmatrix} d\omega',$$

and similarly for probabilities conditioned on GC. If (AT, w_0) initially has probability $p(w_0) f(w_0)$, this becomes, on integration over w_0,

$$\begin{bmatrix} p_{AT,AT}^{(N)}(w) \\ p_{GC,AT}^{(N)}(w) \end{bmatrix} = \begin{bmatrix} 1 - \alpha(w) & \beta(w) \\ \alpha(w) & 1 - \beta(w) \end{bmatrix} \int W(w \mid w') \begin{bmatrix} p_{AT,AT}^{(N-1)}(w') \\ p_{GC,AT}^{(N-1)}(w') \end{bmatrix} dw'$$

where

$$\begin{bmatrix} p_{AT,AT}^{(0)}(w_0) \\ p_{GC,AT}^{(0)}(w_0) \end{bmatrix} = p(w_0) f(w_0) \begin{pmatrix} 1 \\ 0 \end{pmatrix}.$$

Adding the two rows, we see that

$$\rho^{(N)}(w) = p_{AT,AT}^{(N)}(w) + p_{GC,AT}^{(N)}(w)$$

satisfies

$$\rho^{(N)}(w) = \int W(w \mid w') \rho^{(N-1)}(w') \, dw',$$

$$p^{(0)}(w) = p(w) f(w).$$

This allows us to eliminate $p_{GC,AT}^{(N)}(w)$, obtaining, with $p_{AT,AT}^{(N)}(w)$ abbreviated as $p^{(N)}(w)$,

$$p^{(N)}(w) = [1 - \alpha(w) - \beta(w)] \int W(w \mid w') p^{(N-1)}(w') dw' + \beta(w) \rho^{(N)}(w),$$

$$p^{(0)}(w) = p(w) f(w).$$

It is convenient to choose the parameter w as the value of the probability p that the system is "aiming" at: $\binom{p}{1-p} = \begin{bmatrix} 1-\alpha & \beta \\ \alpha & 1-\beta \end{bmatrix} \binom{p}{1-p}$ or $p = \beta/(\alpha + \beta) = w$, and introduce $\gamma = \alpha + \beta$, the total deviation from independence. Hence

$$\alpha(w) = (1 - w) \gamma(w), \qquad \beta(w) = w \gamma(w).$$

Taking the initial $p(w) = w$, as well, we therefore have

$$p^{(N)}(w) = (1 - \gamma(w)) \int W(w \mid w') p^{(N-1)}(w') \, dw' + w \gamma(w) \rho^{(N)}(w),$$

where

$$\rho^{(N)}(w) = \int W(w \mid w') \rho^{(N-1)}(w') \, dw',$$

$$p^{(0)}(w) = \rho^{(0)}(w) = w f(w),$$

subsequent to which, of course,

$$p_{AT,AT}(N) = \int p^{(N)}(w)\,dw.$$

What we need is the function $\gamma(w)$. In practice, we can find this by taking kilobase-pair windows from the sequence of interest, putting them in bins of common $w \pm \frac{1}{2}\Delta w$ and fitting γ to each bin. In the analysis of Fickett and Tung, it turns out that $1 \le \gamma \le 1.1$, very close to independence, $p_{AT|AT} = p_{AT|GC}$, and hence to the standard HMM. To the extent that $\gamma = 1$ is valid, the statistics degenerate to

$$p_{AT,AT}(N) = \int w\rho^{(N)}(w)\,dw,$$

where

$$\rho^{(N)}(w) = \int W(w \mid w')\,\rho^{(N-1)}(w')\,dw',$$

$$\rho^{(0)}(w) = wf(w).$$

There remains the question of the *diffusion* of w, as represented by $\rho^{(N)}(w)$, the iterated action of $W(w \mid w')$ on $\rho^{(0)}(w) = wf(w)$. Assuming that there is only a small change in w each time, we can set $\rho^{(N)}(w') = \rho^{(N)}(w) + (w' - w)(\partial/\partial w)\rho^{(N)}(w) + \frac{1}{2}(w' - w)^2(\partial^2/\partial w^2)\rho^{(N)}(w) + \cdots +$ in the transition equation. If the transitions are symmetric, $W(w \mid w') = W(w' \mid w)$, then $\int(w' - w)\,W(w \mid w')\,dw' = 0$; hence $\int W(w \mid w')dw' = \int W(w' \mid w)\,dw' = 1$, and we define $\sigma^2(w) = \int(w' - w)^2 W(w \mid w')dw'$. The "dynamics" then takes on the Fokker–Planck form (Feller, 1950, Chap. XIV)

$$\rho^{(N)}(w) - \rho^{(N-1)}(w) = \frac{1}{2}\sigma^2(w)\,\frac{\partial^2}{\partial w^2}\rho^{(N-1)}(w).$$

Finally, regarding the dependence on N as continuous, we can write this as

$$\frac{\partial}{\partial N}\rho(w, N) = \frac{1}{2}\sigma^2(w)\,\frac{\partial^2}{\partial w^2}\rho(w, N),$$

$$\rho(w, 0) = w\,f(w),$$

a standard diffusion equation.

In Fickett and Tung (1992) the assumption is made that $\sigma^2(w) = \sigma^2$ is a constant on the allowed range of w and is obtained by fitting the large N

asymptotic solution to the bin distribution above. For the human genome, this corresponds to $\frac{1}{3} \le w \le \frac{2}{3}$, with $\sigma = 0.0015$, and for *E. coli* to $0.40 \le w \le 0.58$, $\sigma = 0.0025$. Further, $f(w)$ was taken as uniform over the full w interval. The nondegenerate diffusion problem, with full $\gamma(w)$, was then solved by simulation, and the AT, $AT(N)$ correlation coefficient computed. The results for *E. coli* are very good, whereas that for human DNA decays a bit too rapidly, perhaps because of the assumption of w-independent σ^2 or truncated uniform $f(w)$.

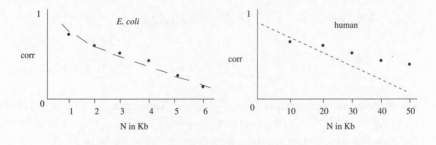

Assignment 5

1. What is the general class of base-symmetric figures of merit $C(n)$? Compute the relevant coefficients in Li's $\bar{C}(n)$.
2. If $\text{Var}^{(N)}(\#u)$ is known as an explicit function of N, find an expression for the required $C_{uu}(n)$.
3. Show explicitly how to specialize the walking Markov model to our version of the Karlin–Brendel patch model.

3.3. Other Measures of Significance

3.2.1. Spectral Analysis

In Section 3.2 we looked at correlations in nominally homogeneous DNA "matter." Now we will start to home in on substructures or meaningful inhomogeneities in pattern. We will soon use the evolutionarily relevant technique of recognizing significance by its presence in more than one species of DNA (or RNA, or protein, or a long segment of DNA) but here we discuss very briefly how we might pick up substructures by internal analysis alone. Again, we need to bunch data in some fashion to reduce noise-driven details, but primitive compressions such as py(CT) or pu(AG) make less sense – they would for example poorly single out things like $(GT)^n$ repeats.

The simplest bunching then consists of continuing to identify bases, and, for a fragment of length N, we define the *Fourier transform*

$$a_k(s) = \frac{1}{N} \sum_{x=1}^{N} e^{2\pi i \, kx/N} \, \delta_x(s),$$

$$\text{integer } |k| < N/2,$$

for each s. Corresponding to this is the *power spectrum*

$$F_k(s)^2 = N \, |a_k(s)|^2 = \frac{1}{N} \sum_{x,y=1}^{N} e^{2\pi i \, k(x-y)/N} \, \delta_x(s) \, \delta_y(s)$$

$$= \frac{1}{N} \sum \delta_x(s) + \frac{1}{N} \sum_{n=1}^{N-1} \left[e^{2\pi i (k/N) n} + e^{-2\pi i (k/N) n} \right] \sum_{1}^{N-n} \delta_x(s) \, \delta_{x+n}(s),$$

which can be amalgamated as $F_k^2 = \sum_s F_k(s)^2$. Noting that $(1/N) \sum \delta_x(s) = p_s$, $(1/N - n) \sum_1^{N-n} \delta_x(s) \delta_{x+n}(s) = C_{ss}(n) + p_s^2$, and $\sum_{n=0}^{N-1}(N-n) \cos 2\pi \, kn/N = 0$ for $k \neq 0$ (show this!), we readily find that, for large N,

$$F_k^2 = 1 - 2 \sum p_s^2 + 2 \sum_{n=1}^{\infty} \left[\cos 2\pi \frac{k}{N} n \sum_n C_{ss}(n) \right], \quad k \neq 0.$$

On the one hand, we have $F_k^2 = 1 - 2 \sum p_s^2$ for random occurrence of bases and $F_k^2 \propto k^{\alpha-2}$ for $\sum_s C_{ss}(n) \sim n^{-\alpha}$ (show this too!) if fractal correlations are indeed a real phenomenon.

On the other hand, F_k^2 is ideal for picking up repeating substructures. Suppose that a repeated pattern P is described by $\delta_{X+x}(s) = v_x$, $x = 1, \ldots, p$ for several values of X; then, for each one, there is a contribution to $N \, a_k(s)$ of

$$\sum_{x=1}^{p} e^{2\pi \, ik(X+x)/N} \, \delta_{X+x}(s) = e^{2\pi \, ikX/N} \, f_k(P),$$

where

$$f_k(P) = \sum_{1}^{p} e^{2\pi \, i \, kx/N} \, v_x$$

is the *form factor* for the pattern P. If the pattern is effectively random, we expect $|f_k(P)|^2 \sim p$. Now if the $\{I_j\}$, $j = 1, \ldots, q$, are the intervals between occurrences of P, so that

$$X_j = \sum_{1}^{j} I_i,$$

the coefficient of $f_k(P)$ in $a_k(s)$ will be

$$b_k(q) = \frac{1}{N} \sum_{j=1}^{q} e^{2\pi i(k/N)\sum_1^j I_i}.$$

If the intervals are estimated as independent and identically distributed, with mean $\bar{I}$ and variance σ^2 (but I_1 is just a half-interval, as is I_{q+1}), then $N = \sum_1^{q+1} I_j = q\bar{I}$, and

$$b_k(q) \sim \frac{1}{N} \sum_{j=1}^{q} \langle e^{2\pi ik/N\, I} \rangle^{j-1/2} \sim \frac{1}{N} \sum_{j=1}^{q} \left(e^{2\pi i\, k/N\, \bar{I} - 2\pi^2 k^2/N^2\sigma^2} \right)^{j-1/2}.$$

Hence if k is any multiple of q, $k = rq = rN/\bar{I}$, so that $e^{2\pi i(k/N)\bar{I}} = 1$, we have

$$b_{rq}(q) \sim \frac{1}{N} \sum_{j=1}^{q} e^{-(j-1/2)\, 2\pi^2 r^2\sigma^2/\bar{I}^2}$$

$$= \frac{1}{N} \left(1 - e^{-2\pi^2 r^3 N\sigma^2/\bar{I}^3} \right)/2 \sinh(\pi^2 r^2 \sigma^2/\bar{I}^2).$$

For σ small, $\sigma < \bar{I}/r$, this just gives $b_{rq}(q) \sim 1/\bar{I}$, a contribution of $\sim pN/\bar{I}^2$ to F_k^2, which can be way above noise. As an example, the 128-bp AT-rich spacer region of Xenopus 5S DNA is shown: The strong peak at $k = \pm 16$ is strong evidence of an eight-base repeated motif, and indeed the observed amplitude is several units of $pN/\bar{I}^2 = 4$. Of course, if $r\sigma/\bar{I}$ is not small,

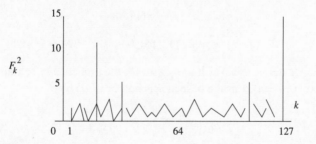

$b_{rq}(q)$ sinks rapidly, and at higher k, the random $|f_k(P)|^2 \sim p$ is approached (see $k = 32$ in the above figure).

For coding regions, the third base of a triplet is a reflection mainly of the mean base frequency in the full sequence, and so one might expect a significant $k = 3$ peak for any variant distribution; this observation has in fact been used to detect genes. The trigonometric functions in the Fourier transform are not especially appropriate for functions that are always small integers, e.g., 0 or 1 for occupancy by a base. More suitable are the Walsh functions, essentially discrete sines and cosines; see, e.g., Tavare and Giddings, 1989. The first five Walsh functions are shown, defined on the interval $[0, 1)$ more generally by

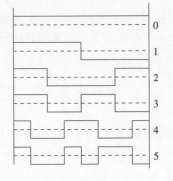

$$W_0(x) = 1, \; x \in [0, 1), \quad W_1(x) = \left. \begin{array}{r} 1 \\ -1 \end{array} \right\} \begin{array}{l} x \in \left[0, \frac{1}{2}\right) \\ x \in \left[\frac{1}{2}, \; 1\right) \end{array},$$

$$W_{2n}(x) = \left. \begin{array}{c} W_n(2x) \\ (-1)^n \, W_n(2x - 1) \end{array} \right\} \begin{array}{l} x \in \left[0, \frac{1}{2}\right) \\ x \in \left[\frac{1}{2}, \; 1\right) \end{array},$$

$$W_{2n+1}(x) = \left. \begin{array}{c} W_n(2x) \\ (-1)^{n+1} \, W_n(2x - 1) \end{array} \right\} \begin{array}{l} x \in \left[0, \frac{1}{2}\right) \\ x \in \left[\frac{1}{2}, \; 1\right) \end{array}.$$

For discrete x as well, say $x = j/N$, where $N = 2^p$, we set

$$W_k(j/N) = w(k, j) \quad \text{for } 0 \le j, k < N = 2^p,$$

and then we define the Walsh transform for the sequence s_j belonging to the base α (e.g., $s = 0, 1$ as the base at j is not, or is, α) by

$$a_k = \frac{1}{N} \sum_0^{N-1} w(k, j) s_j.$$

If this is really to be the analog of the Fourier transform, we will want the inverse to take the same form:

$$s_j = \sum_0^{N-1} w(j, k) a_k.$$

To prove this, we need orthonormality. For this purpose, the representation

$$w(k, j) = (-1)^{\sum_{r=0}^{p-1} j_r(k_{p-r} + k_{p-r-1})} = w(j, k),$$

where

$$j = \sum_0^{p-1} j_r 2^r, \quad k = \sum_0^{p-1} k_r 2^r, \quad (j_r, k_r = 0, 1),$$

is used, which is readily seen to satisfy the preceding recursion relations. Then

$$\sum_j w(k, j) \, w(k', j) = \sum_{\{j_r\}} (-1)^{\sum_r j_r(k_{p-r} + k_{p-r-1} + k'_{p-r} + k'_{p-r-1})}$$

$$= \prod_r \sum_{j=0,1} (-1)^{j(k_{p-r} + k_{p-r-1} + k'_{p-r} + k'_{p-r-1})}$$

$$= 2^p \prod_r \delta_{\mathrm{mod}\,2}(k_{p-r} + k_{p-r-1} + k'_{p-r} + k'_{p-r-1})$$

$$= N \prod_r \delta_{k_r, k'_r} = N \, \delta_{k, k'},$$

which is all we need.

At this point, we can proceed exactly as in the Fourier case, first defining

$$a_k(\alpha) = \frac{1}{N} \sum_0^{N-1} w(k, j) \, \delta_j(\alpha)$$

for each base and noting that $\sum_\alpha \delta_j(\alpha) = 1$ implies that $\sum_{\alpha=1}^4 a_k(\alpha) = \delta_{k,0}$, and also that $a_0(\alpha)$ is the frequency p_α. Then comes the power spectrum:

$$F_k^2 = N \sum_\alpha a_k(\alpha)^2.$$

This is a bit harder to interpret, but in the same example as before, period 8 certainly

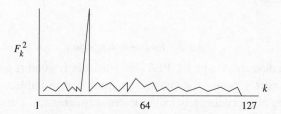

comes up as a very sharp peak with respect to k. In practice, a combination of Fourier and Walsh produces a signature that the expert can easily interpret, but scarcely uniquely.

An analytic technique designed to be particularly sensitive to structure at a given spatial scale in a hierarchical set of scales is that of wavelet analysis (see, e.g., Meyer, 1992, for a high-level treatment). For a prototype, we might imagine replacing the transform kernel e^{ikx} in a Fourier transform with $e^{ikx} e^{-\alpha(k)(x-X)^2}$, creating a "local" Fourier transform that depends on both frequency k and location X. Most (continuous) wavelet transforms that are used take the form

$$\tilde{f}(k, X) = k^{1/2} \int \psi (kx - X) f(x)dx$$

for special ψ, and the $\{k, X\}$ are arranged, typically with $k = 2^j$, such that the $\psi_{kX(x)} = k^{1/2} \psi(kx - X)$ form an orthnormal set of functions, leading to the usual sorts of expansion theorems. The $\tilde{f}(k, X)$ are useful descriptive parameters even if the ψ_{kX} are not orthonormal, and of course extend at once to discrete variables. For a typical application to DNA, see Tsonis et al. (1996).

The task of discovering the words and the phrases of DNA, not to mention the grammar, from internal – one-sequence – indications alone is not difficult in one respect: There are a number of unequally spaced exact and inexact repeats of long (words?) and very long (phrases?) identifiable subsequences. For example, there are Alu, 300 bp, 5% of the human genome, or L1, very long, 4% of the human genome. These are transposable elements, RNA mediated with reverse transcription, and are sufficient, e.g., to identify human DNA in a nonhuman cell. There are also shorter frequent motifs or "sites," picked up out of noise, e.g., by the r-scan technique of Karlin et al. (1989), in which the statistics of distances between identical or almost identical subsequences, spaced r subsequences apart, is examined for significance. However, most of the progress made in finding the longer words of DNA, RNA, and proteins, has come from comparisons of sequences, and it is to this that we will soon turn our attention.

3.2.2. Entropic Criteria

Spectral methods (see, e.g., Li, 1997) are particularly good at picking up repeats and near repeats; the human genome, for example, is estimated as consisting of approximately half repeated subsequences. These repeats, as has been noted, come in many flavors – aside from the obvious codon repeats in exons – from small tandem repeats, through SINES (short interspersed repeat sequences, $10^2 - 10^3$ bp) and on to LINES (long interspread repeat sequences, $\sim 10^4$ bp). There are effective algorithms (Apostolico, 2000) for detecting repeats without the impossibly laborious checking of sequence after sequence. Also, repeats are but one example of anomalously represented subsequences; underrepresented ones occur as well. The existence of a core of strictly avoided words in fact sets up a complex structure of avoidance, which has recently been investigated (Hao et al., 2000). For a survey of the statistics of such events under prototypical random placement, see Karlin et al. (1989). Of course, repeats are not the only special constructs found in the DNA "language"—reverse palindromes are frequent, and "syllables" that are only found together are common as well (Trifonov, 1988).

Thus the DNA language is far from random, both in its coding and in its noncoding subsequences, and a first estimate of where the information resides might proceed by finding the information content, the negative entropy, of selected segments. The entropy per site,

$$H_1 = -\sum_{s=1}^{4} p_s \log_2 p_s,$$

where p_s is the relative frequency of base s, is a very primitive indicator. Clearly $0 \le H_1 \le 2$, with $H_1 = 2$ for the purely random $p_s = 1/4$, and even GC-wide regions can reduce this substantially, but for sizable genomes, $H_1 \sim 1.9 - 2.0$ is universal. More informatively, we can look at words w of length n, occurring at frequency $p_w^{(n)}$, and construct

$$H_n = \sum_{w} -p_w^{(n)} \log_2 p_w^{(n)}.$$

The $n = 3$ codons in coding regions typically produce $H_3 \sim 5.9$, close to the purely random 6, but at large n a better indicator is excess entropy:

$$h_n = H_{n+1} - H_n.$$

In fact, it is seen at once that

$$h_n = \sum_{w,s} -p_{w \cdot s}^{(n+1)} \log_2 \left[p_{w \cdot s}^{(n+1)} / p_w^{(n)} \right]$$
$$= \langle \log_2 \Pr \left[s \mid w^{(n)} \right] \rangle,$$

averaged over n words and the following base, s, so that h_n is a measure of the unpredictability of the next base that is added. More sophisticated definitions of excess entropy (Loewenstern and Ylanilos, 1999) indeed show a substantial reduction, to $h_n \sim 1.6$ for a variety of genomes.

A relatively unbiased measure of excess entropy is obtained from

$$h = \lim_{n \to ``\infty"} h_n,$$

(where "∞" is a lot smaller than the genome length L), as h_n would in fact saturate at h_m for an m-memory Markov chain. The full dependence of h_n on n is a better indicator (Herzl et al., 1994) of a repetitive structure. We can see this quickly without going into great detail, but by focusing on n words that can overlap the left edge of a specific length l repeat R by k bases, where $1 \leq k \leq n - 1$. Suppose the repetition rate is ρ per site, so that in length L there are ρL repeats. Then, because k bases are fixed, the expected number of occurences of the word, i.e., of matches on the genome, is clearly $\rho L / 4^{n-k}$ for those that overlap the left edge of an R by k bases and $L'/4^n$ for those that do not and hence are free. $L' \sim L$ is the number of slots for n words that do not overlap an R. Thus the frequency of such a word is $\rho / 4^{n-k} + 1/4^n$. Now if we append a base to the right end of the word, only one base b is allowed when the word does overlap, but the expected number of $(n + 1)$-word matches in the rest of the genome is now $L'/4^{n+1}$. Hence the entropy contribution to h_n is

$$- \log_2 \Pr (b \mid w) = \log_2 \left(4^{\frac{\rho}{n-k}} + \frac{1}{4^n} \right) \Big/ \left(\frac{\rho}{4^{n-k}} + \frac{1}{4^{n+1}} \right)$$

$$= 2 - \log_2 \left(1 + 4^{k+1} \rho / 1 + 4^k \rho \right),$$

which changes from 2 at small k to 0 at large k, a rapid change occurring for $k_c \sim \log_4 1/\rho$. So if $n < k_c$, the n dependence will be smooth, but as n passes through k_c, a rapid shift can occur, giving us an estimate of the frequency ρ.

4

Sequence Comparison

Having examined aspects of the general structure of the language of DNA, we continue to home in on the words, phrases, etc., of the language. A word of course is a subsequence that recurs in the *same* or *another* organism, either exactly, or distorted, or in synonym form; a phrase consists of key words together perhaps with filler. We will save for Section 4.3 the question of how potential words or phrases are located in the first place in our new more general context, and concentrate now on the degree of confidence with which we can assert that these objects have indeed been found. It must be emphasized that we are only attending to linear ordering, the *primary structure* of the molecule, so that proteins, for example, correspond simply to coding linear subsequences; the fashion in which the distinctive three-dimensional structure arises, clearly crucial for proteins and hardly irrelevant for DNA, is not being addressed.

4.1. Basic Matching

The prototypical situation to be analyzed is this: Two linear chains of length l (nucleotides, amino acids, ... ,), when aligned, are found to have a common (contiguous) subsequence of r units. What is the probability that this was a random event and not an indicator of a functional or an evolutionary relationship between the chains? At the most primitive level of resolution, random means independent selection of the units at the overall frequencies, p_α for the αth type of unit, $\alpha = 1, \ldots, n$ ($n = 4, 20, \ldots$). In this case, essentially the same as that of Subsection 3.2.1, the probability of a match at a given location will be

$$p = \sum_{\alpha=1}^{n} p_\alpha^2,$$

and the probability of not matching will be $q = 1 - p$. Note that the term "match" depends very much on the way equivalent units are defined; the

81

20 amino acids, for example, can for most purposes be simplified to four or five equivalence classes.

4.1.1. Mutual-Exclusion Model

Our precise question will be: what is the probability P, under the assumption of randomness, that a match of length $\geq r$ will occur? Then, if this event occurs, randomness can be rejected in favor of significance at a confidence level of $1 - P$. The key to our present analysis (Percus and Percus, 1994) is that the result will be interesting precisely when P is small, so that any of the probabilities of events of which it is composed will be very small. This means that although the different matches that can occur are not mutually exclusive, i.e., if $(A \text{ and } B) \equiv A \cdot B$, we have $P(A \cdot B) \neq 0$; nonetheless $P(A \cdot B)$ will be second order in smallness, and we can act as if

$$P(A \cup B) = P(A) + P(B),$$

i.e., just add the component probabilities.

Back to $\geq r$ matches ("successes") out of l: the possibilities are that

1. at least r matches in sequence (we term this an r match) start at the left end, probability p^r,
2. an r match starts at one of $s = 2, 3, \ldots l + 1 - r$; hence a mismatch at location $s - 1$ is followed by r matches, probability qp^r.

The single events of type 1. and $l - r$ of type 2. give us, under the assumption of additivity,

$$P^l_{\geq r} = p^r + (l - r)qp^r$$
$$= [1 + (l - r)q]p^r \quad \text{for } r \geq 1.$$

Example: At 0.999 confidence level, a tail probability of $P = 0.001$, and with $p = 1/4$, a match of $r \geq \sim \ln(0.75\,l/10^{-3})/\ln 4 = \ln_4 750l$, e.g., $r \geq 15$ for $l = 10^6$, would be significant.

Now, more realistically, we consider chains of lengths $l_2 \leq l_1$ and ask for a match of $\geq r$ units in a row in one chain with $\geq r$ in a row someplace along the other, i.e., not necessarily in register. We do this by sliding one chain over the other, and we treat the length in common as a pair of chains in register, to which we can apply the above result. We have the following possibilities:

1. The l_2 chain fits inside the l_1 chain at $l_1 + 1 - l_2$ different starting places, each contributing $[1 + (l_2 - r)q]p^r$.

2. The l_1 chain starts to the left of l_2, with $l < l_2$ (and $l \geq r$) sites overlapping, a contribution of $[1 + (l - r)q]p^r$.

3. 2 is repeated on the right. Hence

$$P^{l_1,l_2}_{\geq r} = (l_1 + 1 - l_2)[1 + (l_2 - r)q]\, p^r + 2 \sum_{l=r}^{l_2-1}[1 + (l - r)q]\, p^r$$

$$= [1 + (l_1 - r) + (l_2 - r) + (l_1 - r)(l_2 - r)q]\, p^r, \quad 1 \leq r < l_2.$$

Example: Two DNA fragments, of lengths $l_1 = 154$ and $l_2 = 103$, with nominal $p = 1/4$, would have, if exactly computed, $P_{\geq 9} = 0.039$. The above estimate yields $P_{\geq 4} = 0.040$, which is very close.

Of course, some substitutions can be tolerated without changing functionality of biological subsequences. Suppose for example that the matching frame is $\geq r$, but only $m < r$ sites have to match, including as well the first and the last that define r. Now the elementary composite-match probability, instead of being p^r, is determined by one match, $m - 2$ out of $r - 2$ matches, then another match, a probability of $p\binom{r-2}{m-2}p^{m-2}q^{r-m}p$. We conclude that the only effect is the replacement:

$$\text{for } m \text{ out of } r, \text{ replace: } p^r \to \binom{r-2}{m-2}p^m q^{r-m}.$$

More reasonable might be that at least m out of r match, leading to

$$\text{for } \geq m \text{ out of } r, \text{ replace: } p^r \to \sum_{k=m}^{r}\binom{r-2}{m-2}p^k q^{r-k}$$

The assumption that all successful r matches are "space filling," and therefore mutually exclusive, is clearly an assumption that holds only if the elementary events have sufficiently small probabilities that the mutual probabilities are fully negligible. If we are not sure, some estimate is needed. The simplest goes like this: Consider the basic matching problem. If A_j *is the event that a match, at* $\geq r$ *contiguous sites, starts at* j, for $j = 1, \ldots, l + 1 - r$, then

$$P^l_{\geq r} = P(A_1 \cup A_2 \cup \cdots \cup A_{l+1-r}).$$

However, now we can use the first of the Benferroni inequalities (see, e.g., Feller, 1950, p. 100). These are most easily obtained in terms of Venn diagrams on the space of elementary events. Representing the composite events $A, B, C, \ldots$, by circles and simply counting, we have $P(A \cup B) =$

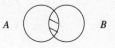

$A \qquad \qquad B$

$P(A) + P(B) - P(A \cdot B)$, so that

$$P(A) + P(B) \geq P(A \cup B) = P(A) + P(B) - P(A \cdot B)$$

At the next stage,

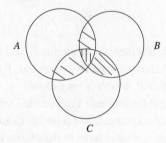

$$P(A \cup B \cup C) = P(A) + P(B) + P(C) - P(A \cdot B) - P(A \cdot C)$$
$$- P(B \cdot C) + P(A \cdot B \cdot C),$$

so that

$$P(A) + P(B) + P(C) \geq P(A \cup B \cup C) \geq P(A) + P(B) + P(C)$$
$$- P(A \cdot B) - P(A \cdot C) - P(B \cdot C),$$

and, more generally (Frechet, 1940),

$$\sum_{j=1}^{l+1-r} P(A_j) \geq P\left(\bigcup_{j=1}^{l+1-r} A_j\right)$$

$$\geq \sum_{j=1}^{l+1-r} P(A_j) - \sum_{1 \leq j < k \leq l+1-r} P(A_j \cdot A_k).$$

Applying this to the basic matching problem, we have $P(A_1) = p^r$, and $P(A_j) = qp^r$ for $j > 1$. If $j < k \leq j + r$, then A_k overlaps or abuts A_j, so that k cannot be the *start* of an r match, and $P(A_j \cdot A_k) = 0$. If $k > j + r$, A_j and A_k are independent, so $P(A_j \cdot A_k) = P(A_j) P(A_k)$. Thus the nonvanishing contributions to the correction come from $j = 1, r + 1 < k \leq l + 1 - r$, a total of $(l - 2r)p^r qp^r$, and $1 < j \leq l - 2r, r_j < k \leq l + 1 - r$, a total of $\sum_{j=2}^{l-2r} (l + 1 - 2r - j)(qp^r)^2 = \frac{1}{2}(l - 2r + 1)(l - 2r)(qp^r)^2$. We conclude that (see, e.g., Uspensky, 1937)

$$[1 + (l - r)q]p^r \geq P_{\geq r}^l \geq [1 + (l - r)q] p^r$$

$$- (l - 2r)\left[1 + \frac{q}{2}(l - 2r - 1)\right]qp^{2r}.$$

Example: For a "fair" coin, $p = \frac{1}{2}$, tossed 2059 times, the exact result to four places is $P_{\geq 13} = 0.1176$. From our inequality, $0.1249 \geq P_{\geq 13}^{2059} \geq 0.1172$, indicating as well the rapid convergence of the alternating series of which we have used only the first two terms.

The general result is that, for the first-order error, $\Delta P \sim P^2$. However, even more importantly, the extremely simple *upper bound* that we get with the mutual-exclusion model is precisely what we need to ensure nonrandomness at a given level of confidence.

4.1.2. Independence Model

A different kind of approximation assumes that successful r matches are independent events, as indeed most of them are. Then $1 - P = P(\bar{A}_1 \cdot \bar{A}_2 \cdots) \sim \prod_i P(\bar{A}_i) = \prod_i [1 - P(A_i)]$ or, for small $P(A_i)$,

$$1 - P \sim \exp - \sum_i P(A_i)$$

in Poisson distribution form. Here, too, the corrections can be arranged to supply bounds on the error made. Let us introduce the variable $X_i = (0, 1)$, with $\bar{X}_i = 1 - X_i$, and use the notation

$$A_i \text{ occurs} \Leftrightarrow X_i = 1,$$
$$A_i \text{ does not occur} \Leftrightarrow X_i = 0.$$

We can then, quite generally, proceed as in Subsection 3.2.2, but we shall do so with a little more control. We want to find $1 - P = \Pr(\prod \bar{A}_i) = E[\prod_i (1 - X_i)]$, but instead we look at

$$I(\gamma) = \ln \left\langle \prod_i (1 - \gamma X_i) \right\rangle.$$

Now

$$I'(\gamma) = -\left\langle \sum_i X_i \prod_{j \neq i} (1 - \gamma X_j) \right\rangle \bigg/ \left\langle \prod_j (1 - \gamma X_j) \right\rangle,$$

$$I''(\gamma) = \left\langle \sum_{i \neq j} X_i X_j \prod_{k \neq i, j} (1 - \gamma X_k) \bigg/ \prod_k (1 - \gamma X_k) \right\rangle$$
$$- \left[\left\langle \sum_i X_i \prod_{k \neq i} (1 - \gamma X_k) \right\rangle \bigg/ \left\langle \prod_k (1 - \gamma X_k) \right\rangle \right]^2.$$

Clearly,

$$I(0) = 0, \qquad I'(0) = -\sum_i \langle X_i \rangle,$$

$$I''(0) = \left\langle \sum_{i \neq j} X_i \, X_j \right\rangle - \left\langle \sum_i X_i \right\rangle^2$$

$$= \sum_{i \neq j} \langle (X_i - \langle X_i \rangle)(X_j - \langle X_j \rangle) \rangle - \sum_i \langle X_i \rangle^2,$$

so that if

$$\lambda = \sum_i \langle X_i \rangle,$$

a Taylor series expansion gives

$$P = 1 - e^{I(1)} = 1 - e^{-\lambda + (1/2)I''(0)\dots}$$

$$= 1 - e^{-\lambda} - \frac{1}{2}\, e^{-\lambda} I''(0) \cdots.$$

Still being very general, suppose there is a distance $d(i, j)$ defined with respect to indices such that A_i and A_j are independent for $d(i, j) > r$. Then $I''(0)$ simplifies to

$$I''(0) = \sum_{\substack{d(i,j) \leq r}}^{i \neq j} (\langle X_i \, X_j \rangle - \langle X_i \rangle \langle X_j \rangle) - \sum_i \langle X_i \rangle^2$$

$$= b_2 - b_1$$

where

$$b_1 = \sum_{d(i,j) \leq r} \langle X_i \rangle, \langle X_j \rangle, \qquad b_2 = \sum_{\substack{d(i,j)! \leq r \\ i \neq j}} \langle X_i X_j \rangle.$$

The *Chen–Stein theorem* (Chen, 1975) converts this to a somewhat more conservative but rigorous bound:

$$|P - (1 - e^{-\lambda})| \leq \frac{1 - e^{-\lambda}}{\lambda}(b_1 + b_2).$$

In particular, in our typical applications, X_i and X_j will be mutually exclusive for $d(i, j) \leq r$ but $i \neq j$, and so we will have $b_2 = 0$.

Computation for the Chen–Stein estimate (Arratia et al., 1990) makes use of precisely the same information as we needed previously in Subsection 4.1.1. We must make sure that the required independence and mutual exclusivity are satisfied, and we will formalize our previous assertion a bit for this purpose.

Consider first the in-register $\geq r$-match problem (as in coin tossing). Let the independent variables $C_k = 0, 1$ denote failure or success of a match at site k; then X_i for a success run $\geq r$ starting at site i is simply

$$X_i = (1 - C_{i-1}) \prod_{k=1}^{i+r-1} C_i, \qquad (C_0 \equiv 0).$$

Clearly X_i is independent of X_j if $j > i + r$ (they have no C_k in common) whereas $X_i X_j = 0$ if $i < j \leq i+r$ (X_i contains C_{j-1} but X_j contains $1 - C_{j-1}$). Furthermore, $\lambda = \langle \sum X_i \rangle = [1 + (l - r)q]p^r$, as in Subsection 4.1.1, and $b_1 = \sum_{1 \leq i, j \leq l-r} \langle X_i \rangle \langle X_j \rangle = (l - 2r) \; qp^{2r} + \frac{1}{2}(l - 2r)(l - 2r - 1)q^2 p^{2r}$, for which a very good estimate is (show this)

$$b_1 < \frac{2r + 1}{l} \lambda + 2\lambda p^r.$$

Example: Fair coin, $l = 2047$, $p = \frac{1}{2}$, $r = 14$:

$$\left| P_{\geq 14}^{2047} - 0.06059 \right| \leq 6 \times 10^{-5}.$$

Continuing to two sequences, $\{X_k\}$ of length l_1 and $\{Y_j\}$ of length l_2, we define $C_{ij} = 1$ if $X_i = Y_j$ and $p = \langle C_{ij} \rangle$. Now the indices are pairs, and, for an $\geq r$ run starting at (i, j),

$$Y_{ij} = (1 - C_{i-1,j-1}) C_{ij} C_{i+1,j+1} \cdots C_{i+r-1,j+r-1}.$$

Here $d(ij, kl) = \min(|i - k|, |j - l|)$, and again $b_2 = 0$. In the same fashion as before, we find

$$\lambda = [(l_1 + l_2 - 2r + 1) + (l_1 - r)(l_2 - r)q] \, p^r,$$

$$b_1 < \frac{(2r + 1)\lambda^2}{(l_1 - r + 1)(l_2 - r + 1)} + 2\lambda p^r.$$

4.1.3. Direct Asymptotic Evaluation

On the assumption of independent base placement in DNA [extension to low-order Markov changes very little (Karlin et al., 1989)], many matching problems can be solved exactly, in principle, and in rapidly convergent asymptotic series, in practice. Consider once more the prototypical run of $\geq r$ successes in a chain of length l. We have seen in Subsection 3.2.1 that for the complementary probability $\bar{P}_{\geq r}^l$ of having no run of r or more, then decomposing the sequence into successive runs of less than r, separated by failures, there results

$$G(z) = \sum_{l=0}^{\infty} \bar{P}_{\geq r}^l z^l = \sum_{s=0}^{\infty} \sum_{\{l_i=0\}}^{\{r-1\}} (pz)^{l_1} \, qz(pz)^{l_2} \, qz \cdots (pz)^{l_s}$$

$$= [1 - (pz)^r]/[1 - z + qz(pz)^r].$$

The simplest expansion is in $\gamma = p^r$ at fixed l:

$$\bar{P}_{\geq r}^l = \text{coeff } z^l \text{ in } 1 - \gamma z^r/(1 - z + \gamma qz^{r+1})$$

$$= \text{coeff } z^l \text{ in } \frac{1}{1-z} \frac{1 - \gamma z^r}{1 + \gamma z^r qz/1 - z}$$

$$= \text{coeff } z^l \text{ in } \frac{1}{1-z}(1 - \gamma z^r)\left[1 - \gamma z^r \frac{qz}{1-z} + \gamma^2 z^{2r} \frac{(qz)^2}{(1-z)^2} + \cdots\right]$$

$$= \text{coeff } z^l \text{ in } \frac{1}{1-z} - \left[\frac{z^r}{1-z} + q\frac{z^{r+1}}{(1-z)^2}\right]\gamma$$

$$+ \left[\frac{z^{2r+1}}{(1-z)^2}q + \frac{z^{2r+2}}{(1-z)^3}q^2\right]\gamma^2 + \dots,$$

yielding

$$P_{\geq r}^l = 1 - \bar{P}_{\geq r}^l$$

$$= [1 + q(l - r)]\gamma - \left[q(1 - 2r) + q^2 \binom{l - 2r}{2}\right]\gamma^2, \dots,$$

just as before.

However, we can just as easily have more structured matching criteria (Peng et al., 1992). Suppose we demand at least t-match sequences of length $\geq r$ to declare a match. This is almost as easy. Now we "tag" any subsequence of r *or more matches* by a variable ξ in order to recognize it. Then we replace the strict failure combination $\sum_0^{r-1} p^k z^k = (1 - p^r z^r)/(1 - pz)$ with $\sum_0^{r-1} p^k z^k + \xi \sum_r^{\infty} p^k z^k = (1 - p^r z^r + p^r z^r \xi)/(1 - pz)$, giving us instead

$$G(z, \xi) = \frac{1 - p^r z^r + \xi \, p^r z^r}{1 - z + qp^r z^{r+1}(1 - \xi)}.$$

Failing configurations are those in which only the powers $1, \xi, \xi^2, \dots, \xi^{t-1}$ are present. Because

$$(\text{coeff } \xi^0 + \text{coeff } \xi^1 + \cdots + \text{coeff } \xi^{t-1})\, G(z, \xi) = \text{coeff } \xi^{t-1} \frac{G(z, \xi)}{1 - \xi},$$

we conclude that now

$$\bar{P}_{\geq r}^{l,t} = \text{coeff } z^l \xi^{t-1} \text{ in } \frac{1}{1-\xi} \frac{1-\gamma(1-\xi)z^r}{1-z+\gamma q(1-\xi)z^{r+1}}$$

from which

$$P_{\geq r}^{l,t} = \gamma^t q^t \text{ coeff } z^l \text{ in } \frac{(z^{rt+t-1} - pz^{rt+t})/(1-z)}{(1-z+\gamma qz^{r+1})^t},$$

and so, to leading order, which is γ^l, we have

$$P_{\geq r}^{l,t} = \gamma^t q^t \left[\binom{l+1-rt}{t} - p \binom{l-rt}{t} \right] + \cdots.$$

In particular, if $l \gg rt$, then $\binom{l-rt}{t} \sim (l-rt)^t/t!$, $\binom{l+1-rt}{t} \sim (l-rt)^t$ $e^{t/(l-rt)}/t! \sim (l-rt)^t/t!$, and therefore

$$P_{\geq r}^{l,t} \sim \gamma^t q^t (l-rt)^t/t!$$
$$\sim \gamma^t q^t l^t e^{-rt^2/l}/t!.$$

4.1.4. Extreme-Value Technique

When an analysis of the significance of an observed sequence is carried out, it is certainly preferable to focus first on a single criterion and then examine the confidence with which we can assert significance. Such a single criterion, as was mentioned in Subsection 3.2.1, might be an extreme value like the length of the longest match formed, to be assessed by its random reference expected value, variance, and so forth. Now, at fixed l,

$$\Pr(r_{\max} = R) = P_{\geq R} - P_{\geq R+1},$$

Hence

$$E(r_{\max}) = \sum_0^\infty R(P_{\geq R} - P_{\geq R+1})$$
$$= \sum_0^\infty RP_{\geq R} - \sum_1^\infty (R-1)P_{\geq R}$$
$$= \sum_1^\infty P_{\geq R},$$

and, in the same fashion,

$$E\left(r_{\max}^2\right) = \sum_1^\infty (2R-1) P_{\geq R}.$$

The asymptotic analysis of Subsection 4.1.3 is not useful for this purpose, as it yields probabilities much larger than 1 for small r. The Chen–Stein bounds do not suffer from the defect at small λ, and indeed Subsection 4.1.3 can easily be modified to accord with fixed λ, but asymptotic in l. However, especially as criteria become more complex, quicker estimates are useful and can be obtained by extreme-value methods (see, e.g., Waterman, 1995, Chapter 11).

As a prototype, again consider the standard run problem. A run must start following a failure, and for l trials there are $\sim lq$ failures, sharply distributed for large l. Following a failure, length R to the next failure has probability qp^R, and the probability of length $< R$ is $\sum_0^{r-1} qp^r = 1 - p^R$. However, then

$$r_{\max} = \max_{1 \le i \le lq} R_i,$$

counting the lq failures in order. Hence $P_{\ge R} = \Pr(r_{\max} \ge R) = 1 - \Pr(r_{\max} < R) = 1 - \prod_i \Pr(R_i < R) = 1 - \Pr(R_i < R)^{lq} = 1 - (1 - p^R)^{lq} \sim 1 - e^{-qp^R l}$, the familiar Gumbel distribution [Gu58]. Then, converting sums to integrals, we obtain

$$E(r_{\max}) = \int_0^\infty \left(1 - e^{-qp^R l}\right) dR.$$

If we define $u = qp^R l$, then $du/u = -[\ln(1/p)]\, dR$, so $E(r_{\max}) = (1/\ln 1/p) \int_0^{ql} (1 - e^{-u})\, du/u$. For large ql, we rewrite the integral as $\int_0^{ql} \int_0^u (1/u)\, e^{-t} dt\, du = \int_0^{ql} \int_t^{ql} (1/u)\, du\, e^{-t}\, dt = \int_0^{ql} (\ln ql - \ln t) e^{-t}\, dt = \ln ql + , \ldots ,$ so

$$E(r_{\max}) = (\ln ql)/(\ln 1/p).$$

The evaluation of the variance is a bit harder, but it can be shown that

$$\mathrm{Var}\,(r_{\max}) = \frac{\pi^2}{6(\ln 1/p)^2} + \ldots.$$

Now let us match two sequences l_1 and l_2 of the same base frequencies and complicate the matching criterion to lead to tests we will perform later: If compared off register, an r match is now defined as a match of

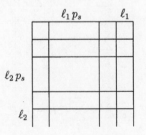

$(1 - \alpha)r$ out of r successive comparisons. Consider base s; there are clearly, on the average, $l_1 p_s \cdot l_2 p_s$ possible matches of base s between the two sequences, and $\sum_s l_1 p_s \cdot l_2 p_s = l_1 l_2 p$ matches of any bases; hence there are $l_1 l_2 - l_1 l_2 p = l_1 l_2 q$ failures in the complete set of comparisons. Again, an r match begins after a failure. We then need $\Pr(r_{\max} \geq R) = \Pr(r_{\text{match}}$ with $r \geq R)$. If the base matches were contiguous, this would be the same as $\Pr(r_{\text{match}}$ with $r = R$, and no restrictions on succeeding sites), and although this is only an approximation for $\alpha \neq 0$, we will use it anyway. Because there are $(1 - \alpha)R$ successes and αR failures required, the latter quantity will be given by $\binom{R}{R\alpha} p^{R-R\alpha} q^{R\alpha}$, so now $\Pr(r_{\max} < R) = \Pr(\text{all} R_i < R) = [1 - \binom{R}{R\alpha} p^{R-R\alpha} q^{R\alpha}]^{l_1 l_2 q} \sim \exp -[l_1 l_2 q \binom{R}{R\alpha} p^{R-R\alpha} q^{R\alpha}]$. We can use the central-limit theorem to approximate the binomial by a Gaussian:

$$\binom{R}{R\alpha} p^{R-R\alpha} q^{R\alpha} \sim \frac{1}{\sqrt{2\pi R p q}} e^{-(R/2pq)(q-\alpha)^2}.$$

Setting $u = l_1 l_2 q (1/\sqrt{2\pi R p q}) e^{-(R/2pq)(q-\alpha)^2}$, we have $du/u = -[(1/2R) + (q - \alpha)^2/(2pq)] dR$, or dropping the $1/\alpha R$ in the dominant large R region, but taking 0 and $l_1 l_2 q$ as the limits for u, we have $E(r_{\max}) = \int_0^\infty \Pr(r_{\max} \geq R) dR = \{(2pq)/[(q - \alpha)^2]\} \int_0^{l_1 l_2 q} (1 - e^{-u}) du/u$, so that

$$E(r_{\max}) = \frac{2pq}{(q - \alpha)^2} \ln l_1 l_2 q + , \ldots,$$

with Var $(r_{\max})$ again $l_1 l_2$ independent.

Two points are to be made. First, the coefficient of $\ln l_1 l_2 q$ diverges as $\alpha \to q$, so that weakening the match criterion leads to the breakdown of the $\ln l$ form. Second, however, the central-limit theorem only applies to a large neighborhood of $q - \alpha$, not including the tail, which dominates for small α; thus it is incorrect for $\alpha = 0$. This is trivially mended: Instead of the normal approximation to the binomial, we just insert Stirling's approximation $x! \sim \sqrt{2\pi x}(x/e)^x$ into the binomial coefficient. This gives us, after a little algebra,

$$\binom{R}{R\alpha} p^{R-R\alpha} q^{R\alpha} \sim \left[\left(\frac{p}{1-\alpha} \right)^{1-\alpha} \left(\frac{q}{\alpha} \right)^\alpha \right]^2 \Big/ \sqrt{2\pi R \alpha(1 - \alpha)}$$

and changes nothing but the coefficient of $\ln l_1 l_2 q$; we now find

$$E(r_{\max}) \sim \ln(l_1 l_2 q) \Big/ \left[(1 - \alpha) \ln \frac{1-\alpha}{p} + \alpha \ln \frac{\alpha}{q} \right] + \ldots.$$

This is indeed correct as $\alpha \to 0$, and coincides with the normal approximation as $\alpha \to q$.

Assignment 6

1. Suppose that base correlations were fractal in form; what effect would this have on the Fourier power spectrum and on the Walsh power spectrum?
2. Apply the mutual-exclusion model to joint $\geq r$ matches for simultaneous comparison of three sequences.
3. For comparison of two sequences, how does amalgamating units explicitly affect the statistics of r_{max}? How does grouping units, i.e., into nonoverlapping doublets or triplets, affect it?

4.2. Matching with Imperfections

4.2.1. Score Distribution

The general procedure in comparing two sequences, protein, DNA, ... , is to algorithmically produce a best alignment and then assess the alignment for significance. If found, we deduce functional similarity, an old example being between platelet growth factor and the v-sis oncogene product, suggesting a growth factor in the latter. Of course, because of ambiguity, replication errors, mutations, evolution, etc., the matches need not be perfect to reflect similarity. We now study how to include this possibility.

When the characterization of an imperfect match is no longer as simple as the length of a sequence of perfect matches (the r, t criterion of A3), it is convenient to define a single quantity, termed a *score*, to epitomize the quality of the match. The (local) score of a two-sequence comparison, $s(A, B)$, will be defined as the maximum score of *aligned* subsequence conformations, i.e., of $s(I, J)$, where $I \subset A$ and $J \subset B$; at this stage, I is required to be a *contiguous* subsequence, as is J. Once we have settled on the basic score $s(I, J)$, the crucial statistical datum for the two-sequence comparison is the cumulative tail probability

$$
\begin{aligned}
\bar{F}(S) &= \Pr[s(A, B) \geq S] \\
&= 1 - \Pr[s(A, B) < S] \\
&= 1 - \Pr[s(I, J) < S, \quad \text{all } (I, J) \subset (A, B)] \\
&= 1 - \left\langle \prod_{(I,J) \subset (A,B)} [1 - X_{IJ}(S)] \right\rangle,
\end{aligned}
$$

where $X_{IJ}(S) = 1$ is the indicator of the event that $s(I, J) \geq S$; otherwise $X_{IJ}(S) = 0$. As a rule, the failure of some intelligently chosen subset denoted by $\{I, J\}$, of the full set of (I, J) is sufficient to imply that of the full set. Now if the corresponding $X_{IJ}(S)$ are small in probability and nearly independent, the effect of the Chen–Stein theorem, and of any number of approximation methods (Goldstein, 1990), is to reduce the preceding probability to the Poisson form:

$$\bar{F}(S) = 1 - \exp\left\{-\sum_{\{I,J\}\subset(A,B)} \Pr[s(I, J) \geq S]\right\}.$$

For this equation to be valid and useful, it is important to select the pairs $\{I, J\}$ to avoid redundancy. This of course does not imply the absence of further collusion between subsets of pairs; the assumption that it does, effectively restricts us to "greedy" match criteria.

In the primitive case of exact matching of units, in which the score is the maximum matching length, we can use the same strategy as in the extreme-value technique of Subsection 4.1.4. Suppose that A and B have lengths l_1 and l_2. Then we choose a pair (i, j) of starting positions of I and J, which are of lengths precisely S, restricted so that $(i - 1, j - 1)$ is not a matching pair. Of the $(l_1 + 1 - S)(l_2 + 1 - S)$ possible pairs, just $q(l_1 + 1 - S)(l_2 + 1 - S)$ on the average, but sharply distributed, will have the desired property. A complete match of I and J satisfies $\Pr[s(I, J) = S] = p^S$, independently of I and J; the corresponding events both exhaust all possibilities of $s(I, J) \geq S$ and are nearly independent of each other. Neglecting S compared with l_1 and l_2, we see that the Gumbel form

$$\bar{F}(S) = 1 - \exp\left(-l_1 l_2 q p^S\right)$$

follows at once.

We can relax the match definition by allowing up to k mismatches in any subsequence comparison, but we retain the score s as the total length of the comparison made; the test is then parameterized by k. With this criterion, we can still proceed by choosing (i, j) to always follow a failure to match, $l_1 l_2 q$ pairs in all, and then impose the condition of $\leq k$ mismatches on the next S units. The probability is $\Pr[s(I, J) = S] = \sum_0^k \binom{S}{m} p^{S-m} q^m$, which again includes all $s \geq S$ in larger subsequence pairs. Although the weak dependence of the I, J matches is further increased, we can still approximate $\bar{F}(S) = 1 - \exp[-l_1 l_2 \sum_0^k \binom{S}{m} p^{S-m} q^{m+1}]$. A fraction α of allowed mismatches proceeds in the same fashion.

The above tests are all characterized by a representation in which the only quantities required are

$$\Pr[s(I, J) = S] = C(S) \rho^S$$

for a selected subclass $\{I, J\}$, and there is no explicit I, J dependence. Here $C(S)$ is a slowly varying function (at least slower than exponential) for large S; ρ is to be computed as

$$\rho = \lim_{S \to \infty} \{\Pr[s(I, J) = S]\}^{1/S},$$

and $C(S)$ as the remaining ratio. If $\Pr[s(I, J) = S] = \sum_{\gamma} a_{\gamma}(S)$ is the sum of a series of at most S^m (positive) terms for fixed m, then $\max_{\gamma} a_{\gamma}(S) \leq \Pr \leq S^m \max_{\gamma} a_{\gamma}(S)$, and we have simply

$$\rho = \lim_{S \to \infty} \left[\max_{\gamma} a_{\gamma}(S) \right]^{1/S}.$$

For an exact match, of course $\rho = p$, $C = 1$; for $\leq k$ mismatches, $\rho = p$ but $C(S) \sim (Sq/p)^k / k!$. For a fraction $\alpha < p/q$ of mismatches, $\rho = (p/1 - \alpha)^{1-\alpha}(q/\alpha)^{\alpha}$ and $C \sim$ const. In all of these cases, we have seen (Subsection 3.1.4) that

$$E(s_{\max}) \sim \ln L / \ln 1/\rho,$$
$$\text{Var}(s_{\max}) \sim (6/\pi^2)/(\ln 1/\rho)^2,$$

where L is the number of IJ pairs required. Of course, this characteristic $\ln L$ dependence is valid only if $\rho < 1$; the $\rho = 1$ situation must always be handled separately.

A more intuitively satisfying way of defining a similarity score is by rewarding matches and penalizing mismatches. The prototype is the linear expression

$$s(I, J) = r(I, J) - \lambda k(I, J),$$

where r is the number of matches, k of mismatches, and the $s_{\max}$ test is parameterized by λ. $\lambda = \infty$ would select only runs of exact matches, whereas $\lambda = 0$ would accept matches no matter how far they are spread apart. We can analyze this very similarly, while realizing that now $s = S$ for a pair I, J no longer implies $s \geq S$ for supersequences. Thus the common length of I and J, say n, is another parameter. After choosing the starting pair (i, j), we must find the probability that $s_{\max} \geq S$ over all (I, J) starting at (i, j). If their common length is n, the probability that $s \geq S$ is simply $\max_n \sum_{\substack{r+k=n \\ r-\lambda k \geq S}} \binom{r+k}{r} p^r q^k$, but there are two regimes to examine. If $\lambda > p/q$, then the maximum of the summand, at $r = np$, $k = nq$, occurs outside of the cross-hatched region $r - \lambda k \geq S$. Thus the maximum term in the sum occurs at $r + k = n$, $r - \lambda k = S$ (to within a fractional remainder) and the probability that $s \geq S$

for *some* value of n satisfies

$$\sum_{r-\lambda k \geq S} \binom{r+k}{r} p^r q^k \geq \Pr(s_{\max} \geq S) \geq \max_{r-\lambda k=S} \binom{r+k}{r} p^r q^k.$$

Clearly then

$$\rho = \lim_{S \to \infty} \Pr(s_{\max} \geq S)^{1/S} = \lim_{S \to \infty} \left[\max_{r-\lambda k=S} \binom{r+k}{r} p^r q^k \right]^{1/S}.$$

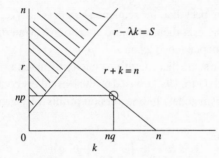

For large λ, k must be small, so observing that $\rho = p$ in the special case $r = S, k = 0$, we see that $\rho \geq p$. For λ not far from p/q, we are close to the maximum of the binomial probability (*at* the maximum if $r, k \to \infty$), which can therefor be approximated by the normal distribution

$$\binom{r+k}{k} p^r q^k \sim [2\pi(r+k)pq]^{-1/2} \exp -\frac{1}{2} [r - p(r+k)]^2 /(r+k)pq.$$

Setting $k = SK$, we have $r = S(1 + \lambda K)$, and we find that this becomes

$$(2\pi \, pq \, S[1 + (\lambda + 1) K]^{-1/2}$$
$$\times \exp -\{(S/2pq)[q + (\lambda q - p) K]^2\}/[1 + (\lambda + 1) K]),$$

whose asymptotic Sth root is $\exp -\{(1/2pq)[q + (\lambda q - p) K]^2 /[1 + (\lambda + 1)K]^2\}$. On maximizing over K, we find

$$\rho \sim \exp - \left[\frac{2}{(1+\lambda)^2 q} \left(\frac{q}{p} \lambda - 1 \right) \right].$$

On the other hand, if λ is large compared with p/q, only $r \sim S$ and $k \sim 0$ contribute to ρ, which hence has the appearance shown.

If $\lambda < p/q$, the maximum in the previous figure is *in* the allowed region, so that $\Pr(s_{\max} \geq S) \sim 1$ for large enough n; (this is because the penalty is small enough that the match sequence will simply become as large as

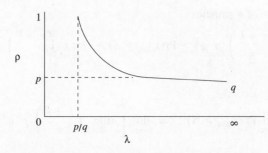

possible), and we expect that, for sequence length $\sim l$, $E(s_{\max}) \sim lp - \lambda lq = lq \, [(p/q) - \lambda]$. There is then a "phase transition" (Waterman et al., 1987) from a $\ln l$ to a cl dependence when $\lambda = p/q$.

A nonparametric score incorporating match/mismatch assessment has been suggested and used (Otto, 1992), based on the least likely contiguous sequence of matches and mismatches. In one version of this very general strategy, we set

$$s = -\ln \begin{pmatrix} r + k \\ r \end{pmatrix} p^r q^k,$$

using the same notation as before, and seek the maximum S over r and k, starting at (i, j). We estimate

$$\Pr[s_{\max}(i, j; r, k) \geq S] \text{ as } \sum_{(r,k) \in I_S} \begin{pmatrix} r + k \\ r \end{pmatrix} p^r q^k,$$

where I_S is the set of (r, k) defined by $\binom{r+k}{r} p^r q^k \leq e^{-S}$, but to compute the parameter ρ, only the maximum term in the sum, e^{-S}, is required. Hence $\rho = 1/e$, $C = 1$, allowing us to conclude, as in Section 3.1, that

$$E(S) \sim \ln l_1 \, l_2, \quad \text{Var}(S) \sim 1.$$

To check the adequacy of the predicted distribution, $F(S) = \exp -(l_1 \, l_2 \, e^{-S})$ for random sequences, a large simulation was carried out, and the distribution of the function $F = F(S)$ plotted. Because $F = \int_0^S \Pr(S') \, dS'$, then we should have $\Pr(S) \, dS = \Pr(F) \, dF = \Pr(F) \, \Pr(S) \, dS$, or $\Pr(F) = 1$ if $F(S)$ is correct. This was verified to high accuracy.

Mismatches are not the only imperfections we should expect. The insertion of bases – e.g., of three-base codons – or deletions may have only a marginal effect on the resulting protein coding or regulatory sequences. These are termed *indels*. For two-sequence comparison, we can allow for either one by inserting blanks in either sequence when testing an alignment and penalize each blank (but not two blanks at the same site, as there was no reason to include them to start with). In other words, although the linear parametric form is not necessarily optimal (Benner et al., 1993), and in particular, the interior indels of a continuous sequence of indels will certainly "cost" less than the first and last indels, we now set

$$s(I, J) = r(I, J) - \lambda k(I, J) - \mu b(I, J),$$

where b is the number of (unpaired) blanks in the alignment. Now, $\mu = \infty$ returns us to the no-indel case and $\lambda = \mu = \infty$ to the perfect-match case, whereas $\lambda = \mu = 0$, caring neither about mismatches nor insertions, should produce very high scores, of the order of the sequence length, with a line of phase transitions in between the extremes. The $\lambda = \mu = 0$ situation, with its anticipated uninformative $E(S) \propto L$, is nonetheless one that we can say something about.

4.2.2. Penalty-Free Limit

Suppose $A = a_1 \cdots a_n$ is a sequence and $S = s_1 \cdots s_m$ is a sequence with $m \leq n$. If both mismatches and insertions are to be ignored, we should *now* define S as a subsequence of A if for *some* $1 \leq i_1 < \cdots < i_m \leq n$, we have $a_{i_k} = s_k$. For sequences A and B, taken for convenience as having a common length n, we then define

$$v(A, B) = \text{max length of subsequence common to } A \text{ and } B,$$

identical with the $\lambda = \mu = 0$ score. We now look for

$$\lim_{n \to \infty} E[v(A, B)]/n,$$

assuming a k-letter alphabet with equal probability of unit selection (Chvatal and Sankoff, 1983; Deken, 1983). Oddly enough, the important conclusion that the limit is bounded from below is easy to show. We simply observe that the maximum occupation of each sequence by some unit must exceed the average, which is n/k, and tag each sequence by its dominant unit, say a. Clearly, any two a sequences have a common subsequence of length $> n/k$. If N_a is the number of a-tagged sequences, then there are N_a^2 pairs with $v(A, B) > n/k$. Hence $E[v(A, B)]/n \geq \sum_{a=1}^{k} N_a^2/N^2 \cdot (1/k)$, where $N = \sum N_a$. However, N^2 is convex, so that $\frac{1}{k} \sum N_a^2 \geq (\frac{1}{k} \sum N_a)^2$. We conclude that

$$E[v(A, B)]/n \geq \frac{1}{k^2} ;$$

of course, it is possible to do better.

It is harder to show that the limit is bounded from above by something less than 1. To do so, we define $F(n, S, k)$ as the number of length n sequences containing a fixed sequence S of length m. We claim that

$$F(n, S, k) = \sum_{j=m}^{n} \binom{n}{j}(k-1)^{n-j}.$$

This is certainly true for $n = 1$, and so we shall carry out induction on n. Note first that the assertion is also true for $m = n$ and for $m = 1$ {the number containing a given unit is the total number, minus the total number lacking that unit, i.e., $k^n - (k-1)^n = [1 + (k-1)]^n - (k-1)^n$, which agrees with the $m = 1$ case}. Then if $1 < m < n$ and $S \subset A$ is a subsequence, we define $A' = a_1 \ldots a_{n-1}$, $S' = s_1 \ldots s_{m-1}$. The set $\{A\}$ is divided into (1) those sites for which $a_n = s_m$, here $S' \subset A'$, so there are $F(n-1, m, k)$ possibilities for A', and (2) sets for which $a_n \neq s_m$, here $S \subset A'$, and a_n is free to have $k - 1$ different values, a total of $(k-1)F(n-1, m, k)$ possibilities. Hence

$$F(n, m, k) = F(n-1, m-1, k) + (k-1)\,F(n-1, m, k),$$

which is indeed satisfied by the asserted formula.

Actually, the expression for $F(n, S, k)$ is stronger than we need. Note that if $j \geq n/k$, then $\binom{n}{j+1}(k-1)^{n-j-1} \leq \binom{n}{j}(k-1)^{n-j}$; it follows that $\binom{n}{j}$ $[(k-1)]^{n-j} \leq \binom{m}{j}(k-1)^{n-m}$ for $j \geq m \geq n/k$, from which

$$F(n, m, k) \leq n\binom{n}{m}(k-1)^{n-m} \quad \text{when } m \geq n/k.$$

The next step, on defining $\theta = m/n > 1/k_i$, is to show that for large n, the proportion $h_k^{(n)}(\theta)$ of pairs (A, B) of length n with $v(A, B) \geq \theta_n$ is bounded by

$$h_k^{(n)}(\theta) \leq H_k(\theta)^{2n}$$

where

$$H_k(\theta) = \frac{h^{\theta/2-1}(k-1)^{1-\theta}}{\theta^\theta(1-\theta)^{1-\theta}}.$$

We do this by "overcounting," i.e., ignoring the fact that the same pair may have more than one common subsequence and just using subsequences to build up pairs. In other words, if $g(n, m, k)$ is the number of pairs with $v(A, B) \geq m$ and $G(n, m, k)$ is the number of triplets (A, B, S) with A and B of length n, S of length $\geq m$, then certainly $g(n, m, k) \leq G(n, m, k)$. It follows that if $\theta > n/k$,

$$h_k^{(n)} \frac{g(n, m, k)}{k^{2n}} \le \frac{G(n, m, k)}{k^{2n}}$$

$$= \sum_s \frac{F(n, s, k)^2}{k^{2n}} \le k^{m-2j} \left[n \binom{n}{m} (k - 1)^{n-m} \right]^2$$

$$= \left\{ k^{\theta/2 - 1} (k - 1)^{1-\theta} \left[n \binom{n}{n\theta} \right]^{1/n} \right\}^{2n}$$

which, by virtue of Stirling's approximation, can be shown to imply the preceding equation.

Now we use this result. First, by examining $(d/d\theta)\ln H_k(\theta)$, we find that the equation $H_k(\theta) = 1$ has a unique solution $\theta = V_k$ in the interval $[1/k, 1]$ and that $H_k(\theta) < 1$ for $\theta > V_k$. Then, for any θ, $V_k < \theta < 1$, we divide the k^{2n} pairs of sequences into two categories: (1) those for which $0 \le v(A, B) \le \theta n$, and (2) those for which $\theta n < v(A, B) \le n$. With the above definition of $h_k^{(n)}(\theta)$ then $E[v(A, B)] < \theta n[1 - h_k^{(n)}(\theta)] + n h_k^{(n)}(\theta) \le n[\theta + H_k(\theta)^{2n}]$; but $H_k(\theta) < 1$, and we can let θ approach V_k. Hence, $\lim_{n \to \infty} E[v(A, B)]/n \le \theta \to V_k$, the desired linear upper bound.

4.2.3. Effect of Indel Penalty

Let us return to the more informative and controllable score for sequences with gaps, $s = r - \lambda k - \mu b$. The needed score distribution is not known exactly, even for the Markov-0 models that we have been considering, but has been approximated on numerous occasions (see, e.g., Zhang and Marr, 1995; Mott and Tribe, 1999; Drasdo et al., 2000). Evolutionary models to assess the confidence with which biological similarity can be inferred from such scores have appeared as well (Hwa and Lassig, 1996), and this is obviously a more realistic direction to take. Here, however, we will attend to a quick and dirty extension of the argument of Subsection 4.2.1 to the statistics of gapped alignment of two thoroughly random sequences. The subsystems being compared now constitute a sequence of matches, mismatches, and unpaired vacancies with respective probabilities p, q, and γ – now $p + q + \gamma = 1$ – and rather than maximize over $n = r + k + b$, we will simply sum:

$$\Pr(s_{\max} \ge S) \cong \sum_{r - \lambda k - \mu b \ge S} \frac{r + k + b!}{r! k! b!} p^r q^k \gamma^b.$$

We can enforce the inequality restriction by inserting a unit step (Heaviside) function $\theta(x) = \int_c e^{itx}/2\pi\, it\, dt$, where c is a real line contour indented below the origin in the complex plan. Hence

$$\Pr(s_{\max} \geq S) \cong \int_c e^{it(s-r+\lambda k+\mu b)} \frac{r+k+b!}{r!k!b!} \, p^r \, q^k \, \gamma^b \, dt/2\pi it$$

$$= \int_c e^{its} \Big/ (1 - pe^{-it} - qe^{i\lambda t} - \gamma e^{i\mu t}) \, dt/2\pi it,$$

where we have used the identity $\sum_{a_1,\dots a_n} [(a_1 + \cdots + a_n!)/(a_1!\dots a_n!)]$
$x_n^{a_1} \dots x_n^{a_n} = (1 - x_1 \dots x_n)^{-1}$.

It follows by a standard argument that if $\rho = e^{it}$ is the root of

$$1 - p/\rho - q\,\rho^\lambda - \gamma\,\rho^\mu = 0$$

of maximum amplitude, then $\lim_{s\to\infty} \Pr(s_{\max} \geq S) = \rho$. For ρ close to 1, we
expand each $\rho^\alpha = \exp \alpha \ln \rho$ to second order in $\ln \rho$, and use $p + q + \gamma = 1$
to obtain

$$\ln \rho = \frac{2}{p + \lambda^2 q + it\,\gamma} \, (p - \lambda q - \gamma\mu).$$

If $\lambda q, \gamma\mu$, and p are of the same order of magnitude and small, we can rewrite
this as

$$-\ln \rho = 2 \left(\lambda \frac{q}{p} + \gamma \frac{\mu}{p} - 1 \right),$$

which indeed reduces properly at the previously analyzed $\mu = 0$ case. When
$\mu \neq 0$, we see that the general structure of the distribution is unchanged,
except for the replacement $\lambda \to \lambda + \mu\,(\gamma/q)$ in the determination of ρ. For
λ and/or μ not small, the original equation is required, and ρ asymptotes to
p. Also when, at small λ, μ, we have

$$\lambda q + \gamma\mu = p,$$

ρ reaches 1 and the linear regime takes over. To be sure, here, as in the case
of the gap-free score, the full distribution function depends as well on the
slowly varying $C(s)$ and the number of relevant pairs $\{I, J\}$.

4.2.4. Score Acquisition

The maximum-score criterion for comparison of sequences A and B is useful
provided that the score $s(I, J), I \subset A, J \subset B$, for an alignment can be rapidly
computed and that alignments can be generated rapidly. At the very least, we
should be able to verify that a score is locally optimal, so that $s(I, J) \geq$
$s(I', J')$ for (I', J') in a suitably defined neighborhood of (I, J). All of this
depends very much on the precise nature of the score that is used. We have seen

that, roughly speaking, scores may be expected to have a behavior with respect to sequence length l that goes between $\ln l$ and cl. Different considerations apply to the two extremes.

Let us first focus on the cl regime, where the constant c has been estimated (Sankoff and Mainville, 1983) for many equiprobable letters, k, in the alphabet – and in the case of mismatch and indel penalty-free criteria – as $2/k^{1/2}$. In fact, the matching criterion can be weakened even further if the test subsequences are allowed to be permuted arbitrarily for the purpose of the test. Hence the maximal subsequence for a sequence A containing $n_1^A, \ldots, n_k^A$ letters of types $1, \ldots, k$, and similarly for B, is represented as $n_1^s = \min(n_1^A, n_1^B), \ldots, n_k^s = \min(n_k^A, n_k^B)$. We can certainly regard $\sum_1^k n_J^s$ as the length of this match, but how do we penalize the remaining letters, $|n_1^A - n_1^B|, \ldots, |n_k^A - n_k^B|$ in number, that do not match with anything? Such a penalty should be nonnegative and vanish only for the identical letter content of the two sequences. Perhaps the simplest is just a weighted rms form

$$d(A, B)^2 = \sum_1^k w_j \left(n_j^A - n_j^B\right)^2,$$

$$\langle n_j \rangle = \frac{1}{2} \left(n_j^A + n_j^B\right),$$

where $w_j = 1/\langle n_j \rangle$ would produce a χ^2 form.

If the letters are bases, the aggregate information estimated by the above $d(A, B)$ – just that of base frequencies – is not a very incisive measure of sequence similarity. However, in the work of Torney et al. (1990), the units are taken as the complete set of $k = 4^m$ words of m nucleotides, or even those between m_L and m_U in length. The predictive power of this score is claimed to be very decent.

Most of the scores traditionally used have been in the $\ln l$ category. Sellers has emphasized (Sellers, 1985; 1986) that these scores tend to rely on the common form

$$s(A, B) = \max_{I \subset A, J \subset B} s(I, J),$$

$$s(I, J) = \rho l(I, J) - d(I, J),$$

$$d(I, J) = \min_{\Lambda} d(I, J, \Lambda),$$

where $l(I, J)$ measures the length of the (I, J) pair (e.g., the mean length of I and J), $d(I, J, \Lambda)$ is a dissimilarity function depending on which alignment Λ is being considered, and ρ is a relative weight. These can be much more strict as to declaring a match, but do have the option of gap insertion in either sequence, the indels previously referred to. How should we organize the possible alignments? A convenient representation, similar in concept to that of Subsection 3.2.1, for the alined gap-inserted subsequences, goes like this: We consider, for example, two subsequences, $I = abccab$ and $J = acbbbc$, aligned as

$$
\begin{array}{lllllllll}
I & a & b & c & c & - & a & b & - \\
\Lambda : \\
J & a & - & - & c & b & b & b & c \, ;
\end{array}
$$

pictorially, we have a vertical line for a gap in sequence J, horizontal for one in sequence I, and diagonal otherwise; each diagonal line can be a match or a mismatch. Sellers sets

$$
d(I, J, \Lambda) = \# \text{ mismatches } + 2 \times \# \text{ gaps},
$$

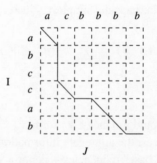

(the first corresponding to mismatches, the second to evolutionary deletions), which is a special case of the Smith–Waterman form of Subsection 3.2.1; in the above alignment, $d(I, J, \Lambda) = 1 + (2 \times 4) = 9$.

With this representation, there is a 1:1 correspondence between (I, J, Λ) alignments and paths (of south, southeast, and east links) from upper left to lower right. There are a number of algorithms for obtaining the dissimilarity $d(I, J) = \min_\Lambda d(I, J, \Lambda)$. The simplest is to define a neighborhood of Λ as any intersecting path; because $d(I, J, \Lambda)$ is additive on subpaths, comparison is easily made, but may result in only a local minimum. There are, however, dynamical programming methods that solve recursion relations for the scores of truncated subsequences. The Needleman–Wunsch (1970) prototype takes the form

$$
\begin{aligned}
d(i, j) = \min\{ & d(i - 1, j) + \Delta(a_i, -), \, d(i - 1, j - 1) + \Delta(a_i, b_j), \\
& \times \, d(i, j - 1) + \Delta(-, b_j) \},
\end{aligned}
$$

where I and J are truncated to their first i and first j elements, respectively

($d = 0$ for negative argument), and Δ is the penalty for the pair or link represented. There is as well a more recent technique (Zhang and Marr, 1995) that uses statistical mechanical methods to compute directly a weighted average of $d(I, J, \Lambda)$ over all Λ.

We now have a global assessment of the match between I and J; the remaining problem is to determine the optimal pair (I, J) of (A, B). Rather than tally all subsequence pairs I, J, we can at least accept (I, J) only if the optimal alignment Λ_0 satisfies $s(I, J, \Lambda_0) = \rho l(I, J) - d(I, J, \Lambda_0) > 0$; then $\rho > d/l$, the mismatch density. Following this, accept only if either (1) for any separation $\Lambda_0 = \Lambda'_0 + \Lambda''_0$, $d(\Lambda_0)/l \le \rho$ as well; or (2) If Λ_0 is overlapped by Λ_1, then $\rho l_0 - d_0 \ge \rho l_1 - d_1$. In fact, a large subclass can be computed by a modified dynamical programming routine; if $s(i, j)$ is the maximum for sequences ending at i, j, it takes the form

$$s(i, j) = \max\{0, s(i - 1, j - 1) + \Delta(a_i, b_j), s(i - 1, j) - \Delta(a_i, -),$$
$$\times s(i, j - 1) - \Delta(-, b_j)\}.$$

However, agreement on the parameters to be used in Δ is not universal.

4.3. Multisequence Comparison

We continue with the question of how to locate and/or characterize sequences with functional similarity mirrored by structural similarity, but now at the multisequence level.

4.3.1. Locating a Common Pattern

Suppose that we have a collection of sequences associated with a common function, such as containing a binding domain for a protein-effecting transcription. How do we characterize this domain, which may very well consist of discernible units – conserved subpieces – but not necessarily with invariant spacings? Presumably there will be an optimal offset of each sequence with respect to the others for lining up the subpieces, but the number of possible off-register arrangements is astronomical. If we have some idea of the length k of the subdomains involved, then one shortcut instead is as follows (Stormo and Hartzell, 1989): We take each sequence, say of common length L, and decompose it into its $L + 1 - k$ (overlapping) constituent words of length k. Now we start with sequence #1, split into $L + 1 - k$ words. We take each word of sequence #2, similarly decomposed, and test it for maximum match among words of #1, append it to its optimal partner in #1, and, in case of a

tie, keep both pairs. Then we test the words of #3 against the set of pairs and
append to the closest pair, and so forth, until the sequences are exhausted.

However, what do we mean by the best match of a word to an n tuplet
of words? A standard assessment is in terms of "information content." If p_s
is the relative frequency of base s in the genome and f_{sj} that of base s at
location j in the whole set of asserted functional fragments, then

$$I_j(f/p) = \sum_1^4 f_{sj} \ln_2(f_{sj}/p_s)$$

is the information content of the set at location j; for a word $j = 1, \ldots, k$,
we correspondingly set $I = \sum_{j=1}^k I_j$. In the preceding procedure, then, the
best matching new word is that which gives the highest information content
when combined with the previous set. After all sequences are combined, we
end up with a set of $\geq L + 1 - k$ *weight matrices* (see also Bucher, 1990),
each being a $4 \times k$ array of f_{sj}, the most informative of which, as in the
figure, signifies a discovered motif. This scheme applies of course to protein
as well as to DNA, but the controlling amino acid interactions depend more
on physical category than specific identity (Miyazawa and Jernigan, 1985),
so that the "threading" techniques (Lathrop and Smith, 1996) used to allocate
sequence to structure to function effectively deal with equivalence classes of
amino acids. Note that, if $f_{sj} \sim p_s$, then a Taylor expansion yields

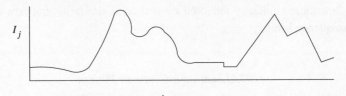

$$I_j \sim \sum_1^4 \frac{1}{2} \frac{(f_{sj} - p_s)^2}{p_s},$$

related to χ^2. Note too that

$$d(f - p) = I\left(\frac{f}{p}\right) + I\left(\frac{p}{f}\right)$$

$$= \sum (f_s - p_s) \ln\left(\frac{f_s}{p_s}\right)$$

is a metric on probability distribution functions, always ≥ 0 and only $= 0$ for
identical distributions.

We can generalize in another way. Suppose that $\{\Lambda\}$ is a putative set of markers asserted to characterize (by presence or absence), e.g., the existence of a type of binding domain, and that these occur at relative frequency $\{f(\Lambda)\}$ in the sample. In "random," modeled by a parameter set $\{\Omega\}$, the model frequencies would be $\{p_\Omega(\Lambda)\}$. Is $\{f(\Lambda)\}$ a significant departure from randomness? In the above, Λ would be taken as a configuration of all bases at all positions in the length k windows. If we imagine the data as the result of $N \to \infty$ trials with independent selection of the $\{\Lambda\}$, then the probability of $\{F(\Lambda) = Nf(\Lambda)\}$ would be $P_N = (N!/\prod_\Lambda F(\Lambda)!)\prod_\Lambda p_\Omega(\Lambda)^{F(\Lambda)}$. The normalized (negative) log likelihood

$$I = \lim_{N\to\infty} -\frac{1}{N}\ln P_N,$$

which, by means of Stirling's approximation, works out to

$$I = \sum_\Lambda f(\Lambda)\ln[f(\Lambda)/p_\Omega(\Lambda)],$$

would then be a legitimate measure of the deviation from randomness. Note that if Λ is decomposed into assertedly independent Λ_i (say the base at site i), so that $p_\Omega(\Lambda) = \prod_i p_\Omega(\Lambda_i)$, and only $f(\Lambda) = \prod_i f(\Lambda_i)$ is computed, then we have instead

$$I = \sum_i \sum_{\Lambda_i} f(\Lambda_i)\ln[f(\Lambda_i)/p_\Omega(\Lambda_i)],$$

which was used in the above.

4.3.2. Assessing Significance

If there is expected to be a common motif, appearing in various mangled forms, but short enough – a minimal functional unit – that insertions and deletions need not be considered, this should be picked up by similarity in – and out of register – multiple occupancy. Because we are looking at a sequence of locations, it is really the significance of an unlikely "run" that is being assessed, and there is no reason why the pair-matching technology should not be extended to this case. Again, let us take the random reference as an independent choice of letters at each site, with probability p_α for type α. Suppose that we have obtained an optimally aligned sample of n sequences. The possibility of a nonlethal mismatch need not be negligible; we therefore define a match at a given location as $\geq k$ of the n sequences exhibiting the same letter. If we observe a contiguous r-site match in the sense just described (extension to t-repeated r matches is direct), then a first estimate of significance would be a comparison with $E(r_{max})$, after which we can fine tune by comparison with the distribution of r_{max}.

The question then is that of $E(r_{max})$ over all (out-of-register) comparisons of sequences of lengths $l_1 \cdots l_n$. Proceeding exactly as we did in the two-sequence comparison, we find that the expected number of sites with at least k out of n matching letters would be the sum over permutations (i.e., *which j are the same*)

$$\sum_{j=k}^{n} \sum_{\text{perm}} \sum_{\alpha} (l_1\, p_\alpha)(l_2\, p_\alpha) \cdots (l_j\, p_\alpha)(l_{j+1}\, q_\alpha) \cdots (l_n\, q_\alpha)$$

$$= \sum_{\alpha} \sum_{j=k}^{n} \binom{n}{j} p_\alpha^j q_\alpha^{n-j}\, l_1 \cdots l_n$$

$$\equiv p\, l_1 \cdots l_n.$$

Again, a run starts after a failure at any of $ql_1 \ldots l_n$ multisequence points, which tells us that

$$E(r_{max}) = \ln_{1/p}(ql_1 \cdots l_n) + \cdots$$

where

$$p = \sum_{\alpha} \sum_{k}^{n} \binom{n}{j} p_\alpha^j q_\alpha^{n-j}.$$

The technical problem of computing p for large n is not trivial and is often solved by large-deviation theory. However, we can be quite direct. The quantity we need is

$$I(x) = \sum_{k}^{n} \binom{n}{j} x^j (1-x)^{n-j},$$

where $k > nx$ in the region of interest, so that a normal approximation is unsuitable. A Poisson approximation can be used, but still better, let us put $I(x)$ in integral form, so that it is completely controllable. To do so, we simply observe that, from $\binom{n}{j}(n-j) = \binom{n}{j+1}(j+1)$, then

$$I'(x) = \sum_{k}^{n} \binom{n}{j} jx^{j-1}(1-x)^{n-j} - \sum_{k}^{n} \binom{n}{j}(n-j)x^j(1-x)^{n-j-1}$$

$$= \sum_{k}^{n} \binom{n}{j} jx^{j-1}(1-x)^{n-j} - \sum_{k+1}^{n} \binom{n}{j} jx^{j-1}(1-x)^{n-j}$$

$$= (n!/k-1!n-k!)x^{k-1}(1-x)^{n-k}.$$

Hence

$$I(x) = \frac{n!}{k-1! \, n-k!} \int_0^x y^{k-1}(1-y)^{n-k} \, dy$$

in terms of the incomplete beta function.

Because the integrand is maximum at $y_0 = k - 1/n - 1$, but $x < y_0$, the integrand rises rapidly as y approaches x, so we can approximate

$$\int_0^x y^{k-1}(1-y)^{n-k} \, dy$$

$$= \int_0^x (x-y)^{k-1}(1-x+y)^{n-k} \, dy$$

$$= x^{k-1}(1-x)^{n-k} \int_0^x \left(1 - \frac{y}{x}\right)^{k-1} \left(1 + \frac{y}{1-x}\right)^{n-k} \, dy$$

$$\sim x^{k-1}(1-x)^{n-k} \int_0^\infty \exp - \left[(k-1)\frac{y}{x} + (n-k)\frac{y}{1-x}\right] dy$$

$$= x^{k-2}(1-x)^{n-k-1} / [(k-1) - (n-1)x].$$

We conclude that p has the very computable form

$$p = \left[\sum_\alpha \frac{p_\alpha^{k-2} q_\alpha^{n-k-1}}{k-1 - p_\alpha(n-1)}\right] k \binom{n}{k}$$

to which, e.g., the Poisson approximation, can also be applied.

Assignment 7

In a model of random evolution, an organism is defined by a string of l composite sites, each one of which can assume one of two forms, labeled by 0 or 1. Transitions from 0 to 1 or 1 to 0 occur at a rate of γ per site per generation, from an initial progenitor. After T generations, two members of the population are compared, according to the score

$$s = r - \lambda k,$$

where r is the number of matches in homologous n-site stretches of the two members, k is the number of mismatches, and s_{max} is the maximum of s over all n.

1. What is the probability that two corresponding sites are the same; different?
2. Find the distribution of s at fixed n, the distribution of s_{max}.

3. Show that the dependence of $E(s_{max})$ on l changes qualitatively when $\lambda = \coth 2 \gamma T$.

4.3.3. Category Analysis

In attempting to understand the language of DNA, we know that there are broad categories of function that constitute the basic structure. These are both in the form of general instructions: splice out the intron that starts here, start transcribing in another 30 bases, ..., and translated properties: this is an exon, this will be involved in phosphorylation, ... In numerous cases, we have a collection of subsequences known to have a common characteristic, and the objective is to find out in what fashion this can be read off as a common characteristic of the base sequence involved, for then the property could be routinely identified from sequence data alone.

An important example is that of splicing signals. Introns are removed from pre-m-RNA by RNA splicing. First the pre-RNA is cleaved at a $5'$-splice site of the intron, separating the sequence into $\cdots$ exon + intron exon ..., and then cleaved at a $3'$-splice site of the intron, with the intron "lariat" removed by a spliceosome (of small ribonuclear protein). A great deal of study on higher eukoryotes has led to the conclusion (Mount, 1982) that *consensus* subsequences

$$5' - \text{exon} - \begin{pmatrix} C \\ A \end{pmatrix} AG\|GT \begin{pmatrix} A \\ G \end{pmatrix} AGT - \text{intron} - 3',$$

$$5' - \text{intron} - \begin{pmatrix} T \\ C \end{pmatrix}_{n \geq 10} N \begin{pmatrix} C \\ T \end{pmatrix} AG\|G - \text{exon} - 3'$$

are involved; here $\begin{pmatrix} C \\ A \end{pmatrix}$ indicates that either C or A is required, N stands for *any* base, and the cut takes place at $\|$. The problem is to decide what combinations of the variable bases indeed give a splice command. A traditional technique is to choose the characteristic being tested for, say identification as a splice signal, construct the corresponding weight matrix $(f_{\alpha j})$ of relative frequencies of base α at site j for a large sample of cases that have been found, and then regard

$$W = \sum f_{\alpha j} \, \delta_j(\alpha)$$

as the score for the sequence being tested, which must exceed some threshold to be evaluated as a "success." Here, again, $\delta_j(\alpha) = 1$ if base α is at site j, else 0. There are a number of intelligent modifications that have been made to accord with presumed composite structure. For example (Shapiro and Senapathy, 1987) in the $3'$-splice problem, there are 10 sites to the left of N

and 4 to the right that appear to be significant. Let h_1 be the highest sum of the highest 8 weights out of 10 in the "training" sample, l_1 the lowest sum of the lowest 8 weights, h_2 the highest sum of 4 weights, l_2 the lowest. Then if t_1 as the highest weight of 8 out of 10 in the test sequence, t_2 the sum of the 4 weights, the score

$$W = \frac{t_1 - l_1}{h_1 - l_1} + \frac{1}{2}\frac{t_2 - l_2}{h_2 - l_2}$$

has empirically proven quite effective.

The above is more than a little bit ad hoc. What we are really trying to do is to separate the sample "points", i.e., subsequence configurations, into two clusters or categories, so that a test "point" can be assessed as to which cluster it is closer to. Quite generally, the nonweighted cluster problem is to separate n data points into c clusters, which can be translated as the following: minimize

$$J(w, c) = \sum_{i=1}^{n} \sum_{a=1}^{c} w_{ia}\, d_{ia}^2$$

subject to

$$\sum_{a=1}^{i} w_{ia} = 1,$$

where $w_{ia} = 1$ if point i is assigned to cluster a and d_{ia} is the Euclidean distance between point i and the centroid of cluster a. This can be solved by "simulated-annealing" minimization techniques.

An alternative approach is by means of classical linear discriminant analysis. Fickett and Tung (1992) looked at the problem of distinguishing between coding (exon) and noncoding segments (introns) of DNA. Among various suggested criteria, they found that dicodon usage – the frequencies of the 4096 different hexamers in a segment – was as effective as any more sophisticated criterion. The formalism produces a coefficient vector such that for a window characterization vector f, the window is "coding" if $c \cdot f > t$ for some threshold t. Here, we determine c by maximizing the ratio of the between-population variation of $c \cdot f$ to the within-population variation of $c \cdot f$. Specifically, suppose that f_{wvj} is the dicodon $-j$ frequency for the wth window observed of class v (coding or noncoding), $\bar{f}_{vj}$, is the mean over w, and $\bar{f}_j$ is the mean over w and v. Then the total covariance matrix T is given by

$$T_{jk} = \sum_{w,v}(f_{wvj} - \bar{f}_j)(f_{wvk} - \bar{f}_k),$$

the within-covariance W by

$$W_{jk} = \sum_{w,v}(f_{wvj} - \bar{f}_{vj})(f_{wvk} - \bar{f}_{vk}),$$

and the between-covariance by $B = T - W$. Maximizing the ratio of variations $c \cdot Bc/c \cdot Wc$ means that c belongs to the maximal eigenvalue λ of B with weight $W : Bc = \lambda Wc$, or

$$W^{-1} Bc = \lambda c,$$

and the threshold t is generally determined so that the fraction of errors on coding windows equals the fraction on noncoding windows.

Computationally, the evaluation of W^{-1} can be a problem, especially because redundancy of information makes W nearly singular. We can avoid this by replacing W with its diagonal part W_D, resulting in what is equivalent to the Penrose discriminant. It is this version that was used, resulting in a 70% accuracy in the above study.

A similar technique has been used (Hayashi, 1952; Iida, 1987, 1988) for splice signal identification; here the $3'$-splice signature was taken as the full 16 nucleotide stretch corresponding to the consensus $\binom{T}{C}_{11}N\binom{T}{C}AG\|G$. The training sample was described by $\delta_j^{(v,w)}(\alpha) = \{^1_0\}$ as base α is or is not present at site j in the wth item from a collection of N_1 splice ($v \equiv 1$) and N_2 nonsplice ($v = 2$) sequences. A linear scor

$$S^{(v,w)} = \sum_{j,\alpha} C_j(\alpha)\, \delta_j^{(v,w)}[\alpha]$$

was set up for the item (v, w), and the problem was expressed as that of finding the 64 weights, not simply as a frequency weight matrix, but so that the $v = 1$ scores are clustered as far from the $v = 2$ scores as possible. Again, the traditional criterion was used: We define

$$\bar{S}^{[v]} = \frac{1}{N_v}\sum_{w} S^{(v,w)}, \quad \bar{S} = \frac{1}{N}\sum_{v,w} S^{(v,w)},$$

$$\sigma_{TOT}^2 = \frac{1}{N}\sum_{v,w}\left(S^{(v,w)} - \bar{S}\right)^2,$$

$$\sigma_{\text{BET}}^2 = \frac{1}{N}\sum_{v} N_v\left(\bar{S}^{(v)} - \bar{S}\right)^2$$

$$= (N_1 N_2/N^2)\left(\bar{S}^{[1]} - \bar{S}^{[2]}\right)^2$$

and maximize $\sigma_{\text{BET}}^2/\sigma_{TOT}^2$, i.e., we maximize the distance between centers of mass relative to the total standard deviation. The $\{C_j(\alpha)\}$ evaluated in this way of course tell us which sites are important and which contribute mainly

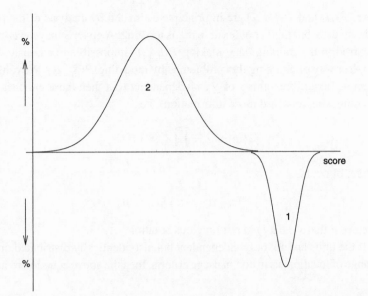

to the noise. The corresponding scores, $S = \sum_{j,\alpha} C_j(\alpha) \delta_j[\alpha]$, then applied to the training data, show a very good separation between type 1 and type 2, yielding greater than 90% selection accuracy, and have been used to predict new splice sites as well

4.3.4. Adaptive Techniques

Bayesian Analysis

There is a more direct estimation of category (see, e.g., Farber et al., 1990; Baldi and Brunak, 1998). Suppose that the datum to be classified consists of a sequence $\underline{S} = \{s_i\}$ of bases, or of codons, or of dicodons, or of amino acids, or, ..., each unit of which may be specified in various redundant ways by a bit sequence. For example, 5 bits for an amino acid or spacer (end of protein) will suffice, but we can also use 21 bits, all of which vanish but one – a typical neural net specification which we also used in the Iida example. If the categories are labeled by μ, Bayes's theorem tells us that the probability that a sequence $\underline{S}$ being tested belongs to μ can be written as

$$P(\mu \mid \underline{S}) = P(\underline{S} \mid \mu)P(\mu)/P(\underline{S})$$

where

$$P(\underline{S}) = \sum_{\nu} P(\underline{S} \mid \nu)P(\nu).$$

Here, $P(\mu)$ and $P(\underline{S} \mid \mu)$ are in principle estimated by analysis of the prior training data, but this is not feasible if $\underline{S}$ is too long: A given $\underline{S}$ may *never* have occurred in the training data, making $P(\underline{S} \mid \mu)$ impossible to estimate. The simplest way of avoiding this problem is by modeling $P(\underline{S} \mid \mu)$. We can, for example, regard the units s of $\underline{S}$ as independent (but then these units should be composite, to afford more information), i.e.,

$$p(\underline{S} \mid \mu) = \prod_i P(s_i \mid \mu),$$

where, of course,

$$P(s_i \mid \mu) = E\left[\delta_i(s_i) \mid \mu\right]$$

(the event that s_i is at i) in our previous notation.

If the units are not only independent but also identically distributed under change of location, as in codon usage criteria, then the above equation reduces to

$$P(\underline{S} \mid \mu) = \prod_s P(s \mid \mu)^{N_s},$$

where N_s is the number of times that s occurs in $S = \{s_i\}$, so that

$$P(\mu \mid \underline{S}) = \frac{P(\mu) \exp \sum_s N_s \ln P(s \mid \mu)}{\sum_v P(v) \exp \sum_s N_s \ln P(s \mid v)}.$$

In particular, for dichotomic μ, i.e., true or false, we can write

$$P(T \mid \underline{S}) = \frac{P(T) \exp \sum_s N_s \ln p_s}{P(T) \exp \sum_s N_s \ln p_s + p(F) \exp \sum_s N_s \ln q_s}$$

$$= \frac{1}{1 + \exp(\sum_s T_s N_s + \theta)},$$

where $p_s = P(s \mid T)$, $q_s = 1 - p_s$, $T_s = \ln q_s/p_s$, and $\theta = \ln[P(F)/P(T)]$, all of which are obtainable from the training set. This sigmoidal form, 0 for large argument, 1 for small – or the reverse – is a standard transformation from quantitative data to approximate yes or no.

A more sophisticated model uses a Markov chain:

$$P(\underline{S} \mid T) = P_1(s_1)P_{12}(s_2 \mid s_1) \cdots P_{N-1,N}(s_N \mid s_{N-1})$$

$$= P_{12}(s_1, s_2)P_{23}(s_2, s_3) \cdots P_{N-1,N}(s_N, s_{N-1})/P_2(s_2) \cdots P_{N-1}(s_{N-1})$$

$$= P_1(s_1) \cdots P_N(s_N) g_{12}(s_1, s_2) \cdots g_{N-1,N}(s_{N-1}, s_N),$$

where $g(s, t) = P(s, t)/P(s) P(t)$ is the neighbor correlation coefficient. For identically distributed singlets and pairs, we have, in obvious notation,

$$P(\underline{S} \mid T) = \prod_s p_s^{N_s} \prod_{s,t} p_{st} \, N_{st},$$

leading, in equally obvious notation, to

$$P(T \mid \underline{S}) = 1/1 + \exp \left(\sum_s T_s \, N_s + \sum_{s,t} T_{st} \, N_{st} + \theta \right).$$

Neural Networks

There is certainly more significant information available in the independently but not identically distributed site model. Suppose that μ is not necessarily dichotomic, and in conformity with computational realizations, let us indeed imagine that we have adopted a binary string representation, with the n possibilities at each composite site represented by a 1 in a substring of n bits, all of the others being 0. Then if $p_j(\mu) = \mathrm{prob}(x_j = 1 \mid \mu)$, we have

$$P(\underline{X} \mid \mu) = \prod_1^N p_j(\mu)^{x_j},$$

so that if $w_{\mu j} = \ln p_j(\mu)$ and $b_\mu = \ln P(\mu)$, we can write

$$P(\mu \mid \underline{X}) = e^{u_{\mu(\underline{X})}} \bigg/ \sum_\nu e^{u_{\nu(\underline{X})}},$$

where

$$u_{\mu(\underline{X})} = \sum_j w_{\mu,j} \, x_j + b_\mu.$$

The problem of course is to use training information to determine the parameters $w_{\mu j}$, b_μ which can thereafter be used for testing mystery data. Once this is done, the μ to be allocated to the data would, e.g., be that which maximizes $P[\mu \mid \underline{X}]$; ideally, the maximum would be close to 1 and the others close to 0. In the end, we will have developed in this way a vector of functions $[f_\mu(\underline{X})]$ whose values will be the type μ probabilities associated with the input data $\underline{X}$. The specific form, $P(\mu \mid \underline{X})$, given above is reasonable but relies on simplifying assumptions to be able to estimate the unknown parameters in the function from limited training data. A class of representation functions, more complex (but hardly general) and with similar components, is that termed "neural networks." Here, the unit is the multiple input–few output function $P(\mu \mid \underline{X})$, which by itself represents all networks of the two-layer "perceptron" architecture. If the units are cascaded by "hidden layers"

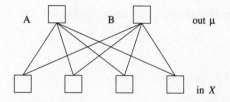

between actual input and output, then it is termed a neural network. In either case, the nonlinearity at any level can be accentuated if any $P(\mu \mid \underline{X})$ is replaced with $e^{\gamma u_\mu(\underline{X})} / \sum_\nu e^{\gamma u_\nu(\underline{X})}$, which as $\gamma \to \infty$ is indeed 1 for the maximum μ, otherwise 0. In strictly binary representation, which is the norm, this is equivalent to $[P(1 \mid \underline{X}), P(0 \mid \underline{X})] = \frac{1}{2}\{1 \pm \operatorname{sgn}[u_1(\underline{X}) - u_0(\underline{X})]\}$, and would certainly be used for the output layer if the training data were used one piece at a time, as its value is then deterministic, not probabilistic.

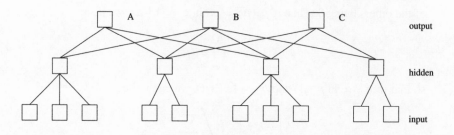

There are then many internal parameters, which we can designate as the set $\{w_\alpha\}$, that enter into the input–output function denoted by $F_\mu(\underline{x}, \underline{w})$. In the process of "training" the network, we want to home in on optimal values of the parameters w_α to best represent the input–output relationship. In practice, some feeling as to how the inputs should be clustered in a qualitative fashion means that many connections will not be made, i.e., the corresponding w are set equal to 0 and do not appear in $F_\mu(\underline{x}, \underline{w})$. Then if training sets $\underline{x}^{(k)}$ are entered, whose output characteristics, say $p_\mu^{(k)}$, are known, the $\underline{w}$ are to be adjusted so that the errors between the $F_\mu(\underline{x}^{(k)}, \underline{w})$ and the $p_\mu^{(k)}$ are minimized. This depends on how we define the error. One definition is simply to sum the mean square error over the whole training set

$$E = \sum_{k, \underline{\mu}} \left[p_{\underline{\mu}}^{(k)} - F_{\underline{\mu}}(\underline{x}^k, \underline{w}) \right]^2$$

and minimize to obtain the optimal $\underline{w}$. Minimization of a many-variable function is a fine art, the most primitive version of which ("conjugate gradient") is an iteration in which a guess $\underline{w}$ is corrected by setting

$$\Delta w_\alpha = -\gamma \, \partial E / \partial \, w_\alpha;$$

if this is really a small change, then $\Delta E = \sum (\partial E / \partial w_\alpha) \Delta w_\alpha = -\gamma \sum (\partial E / \partial w_\alpha)^2 < 0$, as desired. If the uncorrected $\underline{w}$ is the result of the previous training sets and the correction is that which is due to inclusion of further sets, we speak of backpropagating the error to modify the $\underline{w}$. Other techniques, such as simulated annealing, allow a fraction of increases in E as well, to avoid any local-minimum traps.

A simple modification often used is to balance the total weights of the categories in the training set, e.g.,

$$E = \sum_{k,\mu} \left(p_\mu^{(k)} - F_\mu(\underline{x}^{(k)}, \underline{w}) \right)^2 / N_\mu,$$

where N_μ is the number of times that μ occurs in the training set. A rather different form maximizes the mutual information previously introduced. For this purpose, we first need the relative frequency $p_{\mu, \underline{v}}$ of inserting a training pattern characterized by $\underline{v}$ and concluding that the pattern is μ, for a given $\{w_\alpha\}$. The mutual information is then

$$M\{\underline{w}\} = \sum_{\mu, \underline{v}} p_{\mu, \underline{v}} \ln(p_{\mu, \underline{v}} / p_{\mu-} \, p_{-\underline{v}}).$$

Neural networks are not magic; they deal with a very small subset of parameterized output functions of the available input. However, if we have empirical, even anecdotal, information as to the important paths from input data to output characterization, this can be incorporated into the neural network structure (conversely, the optimal set of $\{w_\alpha\}$ determined in the process may give a hint as to the biochemical paths involved – this is certainly the

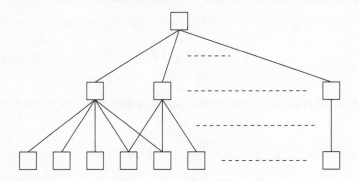

case for the organization of analogous linguistic data in the so-called connec-
tionist form). For example, the prediction of an α helix, a β sheet, or a coil
secondary structure of proteins from the primary amino acid sequence, by
use of the three-output network structure given above (Stolorz et al., 1992),
does no better than 60%–70% accuracy, whereas a corresponding highly
structured network for detecting *E. coli* transcriptional promoters (Abremski
et al., 1991) has close to 100% accuracy in postdiction and is very effective
in prediction. Here, the 4 bases $ACGT$ were given a 4-bit representation, and
the significant information was that of 6 bases from the -10 region, 6 from
the -35 region, which however could be separated from the -10 region by
15–21 bases. Then, in the neural network, the 7 possible overlapping 6 –
base regions, together with the 6 – base region at -10, were represented by
72 input units and connected to 8 hidden units, 1–24 to the first, 5–28 to the
second, etc. The 8 were then connected to 1 output unit. In the perceptron
version, all 72 were connected to the single output. Both versions had the
high accuracy quoted above, showing how important assessment of relevant
information becomes.

Hopfield Networks

In the feed-forward network design, the training of the network requires feedback as well, which we do "by hand" by setting the network parameters to
minimize the difference between output and desired output. More complex
processing of the input could be carried out if feedback existed in the network itself, creating what is termed a Hopfield network; see, e.g., Beale and
Jackson, 1990. The power of a Hopfield network is more evident in a somewhat different, but also somewhat related, application, in which the network,
completely connected, node to node, starts out with a "memory" of many
possible outputs. Given any input, it then proceeds in an unsupervised way to
find the most closely related output.

Specifically, suppose we want to remember a set of patterns

$$\rho^s = \left\{ \sigma_i^s; \quad i = 1, \dots N \right\}, \quad s = 1, \dots, M$$

where

$$\sigma_i = \pm 1.$$

Then an idealized discrete dynamics might update the setting according to

$$\sigma_i' = \operatorname{sgn} \left(\sum_1^N T_{ij} \, \sigma_j \right).$$

How should the node i–node j connection matrix be determined? An obvious choice would be

$$T_{ij} = \frac{1}{M} \sum_{s=1}^{M} \sigma_i^s \sigma_j^s$$

because *if* the patterns were orthogonal,

$$\sum_j \sigma_j^s \sigma_j^t = N \delta_{st},$$

then we would have $\sum T_{ij} \sigma_j^t = (1/M) \sum_{s_{ij}} \sigma_i^s \sigma_j^s \sigma_j^t = (N/M) \sigma_i^t$, and the updating would stop as soon as any pattern ρ^s is reached. In practice, a sizable fraction of M nonorthogonal patterns still serve as attractors of the dynamics.

5

Spatial Structure and Dynamics of DNA

5.1. Thermal Behavior

We have paid little attention to the way in which DNA transmits its information. There are many physicochemical steps involved, with a common theme of energy propagation and localization. Much activity has been devoted to the large-scale behavior of DNA, modeled, e.g., as a belt with elastic properties and giving rise to the "supercoiled" configuration responsible for placing linearly distant sections in close proximity, spatially. The double strand under such motions is stiff ("persistent") at a resolution of some 50 bp. On the other hand, the overt mechanism of information transmission is that of transcription to RNA involving a "transcription bubble" or strand separation of only some 20 bp, activated by RNA polymerase contact. Of course, strand separation should be easier where it is required, and if we heat DNA – as a metaphor for uniform energy transfer – the "melting" is stepwise and nonuniform, presumably a function of the local properties of the (double) molecule. In fact, we might anticipate that such heterogeneous dynamics would appear even in homogeneous DNA, as this would imply greater sensitivity to external stimuli.

Quite different aspects are involved in the dynamics of DNA, RNA, and protein. The first minimally realistic model to be solved along these lines was that of local denaturation (H-bond breaking) of DNA, a requirement for strand separation to allow for transcription (Maddox, 1989, in *Nature*, said in effect that this work was the greatest discovery since the zipper). The model (Techera et al., 1989; Peyrard and Bishop, 1989) is literally that of H-bond connections that lose their integrity when stretched too far by chain vibrations, taken here as thermally excited. As model energy we will choose the ultrasimplified

$$H = \sum_n \left\{ \frac{m}{2} \left(\dot{u}_n^2 + \dot{v}_n^2 \right) + \frac{K}{2} [(u_n - u_{n-1})^2 + (v_n - v_{n-1})^2] + V(u_n - v_n) \right\}$$

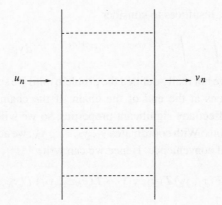

where

$$V(u - v) = D\left[e^{-a(u-v)/\sqrt{2}} - 1\right]^2.$$

Only the out-of-phase $y_n = \frac{1}{\sqrt{2}}(u_n - v_n)$ motions stretch bonds, so we can eliminate the $x_n = (1/\sqrt{2})(u_n + v_n)$, retaining only the y energy:

$$H_y = \sum_n \left[\frac{m}{2}\dot{y}_n^2 + \frac{K}{2}(y_n - y_{n-1})^2 + D(e^{-ay_n} - 1)^2\right].$$

In this reduced model, in which one pays homage to the strong directionality of the hydrogen bond, the twist of the two-strand ladder, and indeed any motion transverse to y_n, becomes irrelevant – but the basic phenomenology is not disturbed.

A first step in the analysis is traditionally that of assuming that the system is in thermal equilibrium, and we accomplish this by asserting that, at reciprocal temperature β, the unnormalized probability of a configuration $(y_1, \ldots, y_N, \dot{y}_1, \ldots, \dot{y}_N)$ is given by the Boltzmann factor

$$\rho(y, \dot{y}) = e^{-\beta H_v(y, \dot{y})}.$$

The basic construct in statistical mechanics is the normalization factor, or *partition function*, subsequent to whose computation all of the system's properties are readily found. Here then, the partition function Ξ is

$$\Xi = \iint e^{-\beta \frac{m}{2} \Sigma_n \dot{y}_n^2}\, d\dot{y}_1 \cdots d\dot{y}_N$$

$$\int \cdots \int e^{-\beta \frac{K}{2}(y_n - y_{n-1})^2}\, e^{-\beta D(e^{-ay_n} - 1)^2}\, dy_1 \cdots dy_N.$$

The velocity integrations separate out, so if we are interested in positional amplitudes alone, it suffices to consider

$$\Xi' = \int \cdots \int e^{-\beta \frac{K}{2}(y_n - y_{n-1})^2} e^{-\beta D(e^{-a y_{n-1}} - 1)^2} \, dy_1 \cdots dy_N.$$

Of course, we have not really defined the problem until we supply information as to what happens at the end of the chain. If the chain is long enough, this should not affect any significant properties, so we will choose periodic boundary conditions: With coordinates $y_1, y_2, \ldots, y_N$, we act as if $y_{N+1} = y_1$, as a mathematical convenience. Hence we can write

$$\Xi' = \int \cdots \int T(y_1, y_2) T(y_2, y_3) \cdots T(y_{N-1}, y_N) T(y_N, y_1) \, dy_1 \cdots dy_N$$

where

$$T(y, y') = \exp -\beta \left[\frac{K}{2}(y - y')^2 + D(e^{-a y} - 1)^2 \right],$$

termed the transition operator, can be regarded as a matrix with continuous indices, so that

$$\Xi' = \operatorname{Tr} T^N.$$

Now what are we looking for? Presumably, for a qualitative change in the distribution of strand separation as temperature is raised. However, the system can be regarded as a set of one-dimensional particles $y_1, \ldots y_N$ under an external potential $V(y)$ and with next-neighbor interaction $\frac{1}{2} K(y_n - y_{n-1})^2$, and a theorem of van Hove assures us that, even in the limit $N \to \infty$ needed to show a sharp phase transition, $(1/N) \ln \Xi'$ and associated physical quantities remain analytic in all parameters. Indeed, if we seek a state in which the y_N are localized – so that we can delocalize them by raising the temperature, this would require a trapped phase in equilibrium with a phase of zero vapor pressure, which certainly will not happen in one dimension. One way of enabling a qualitative change is by a limiting operation, imagining K as arbitrarily large, so that elasticity appears only on a large length scale. Without actually scaling every thing so that the limit can be taken, let us see how this works out.

Leaving the potential $D(e^{-a y} - 1)^2 = U(y)$ unspecified for the moment, we need to carry out the operation

$$Tf(y) = \int T(y, y') f(y') \, dy'$$

$$= e^{-\beta U(y)} \int e^{-\beta K/2(y - y')^2} f(y') \, dy'.$$

Because the Fourier transform $\int e^{iky} \int e^{-\beta K/2(y-y')^2} f(y')\,dy'dy = \int\int e^{ik(y+y')}e^{-\beta/2\,Ky^2} f(y')\,dy\,dy' = (2\pi/\beta K)^{1/2}\,e^{-k^2/2\beta K} \int e^{iky'} f(y')\,dy' = (2\pi/\beta K)^{1/2} \int [e^{\frac{1}{2\beta K}(d/dy')^2} e^{iky'}] f(y')\,dy' = (2\pi/\beta K)^{1/2} \int e^{iky'} e^{\frac{1}{2\beta K}(d/dy')^2} f(y')\,dy'$, we can write

$$Tf(y) = e^{-\beta U(y)} e^{\frac{1}{2\beta K}(d/dy)^2} f(y)(2\pi/\beta K)^{1/2}$$

so that

$$\Xi' = (2\pi/\beta K)^{N/2}\,\mathrm{Tr}\big[e^{-\beta U(y)} e^{\frac{1}{2\beta K}(d/dy)^2}\big]^N$$

$$= (2\pi/\beta K)^{N/2}\,\mathrm{Tr}\big[e^{-\frac{1}{2}\beta U(y)} e^{-\frac{1}{2\beta K}(d/dy)^2} e^{-\frac{1}{2}\beta U(y)}\big]^N.$$

According to a slightly modified Baker–Campbell–Hausdorf expansion (Hausner and Schwartz, 1968) we have

$$e^{A/2}\,e^{B}\,e^{A/2} = \exp\left(A + B + \frac{1}{12}[B,[B,A]] - \frac{1}{24}[A,[A,B]] + \cdots\right)$$

where

$$[A, B] \equiv AB - BA,$$

so

$$e^{-\frac{1}{2}\beta U(y)} e^{\frac{1}{2\beta K}(d/dy)^2} e^{-\frac{1}{2}\beta U(y)}$$

$$= \exp\left(-\beta U(y) + \frac{1}{2\beta K}(d/dy)^2\right.$$

$$-\frac{1}{K}\left\{\beta/24\,U'(y)^2 + 1/(48\beta K)\left[\frac{d^2}{dy^2}\,U''(y)\right.\right.$$

$$\left.\left.\left. + 2\frac{d}{dy}\,U''(y)\,\frac{d}{dy} + U''(y)\frac{d^2}{dy^2}\right]\right\} + \cdots\right).$$

Neglecting the $1/K^2$ terms, we therefore have made the replacement

$$T = (2\pi/\beta K)^{1/2}\exp -\left\{\frac{-1}{2\beta K}\frac{d^2}{dy^2} + \beta[U(y) + U'(y)^2/24K]\right\}.$$

Now it is easily seen that the expectation of y_1 (equivalent to that of any y_j) is given by

$$E(y) = \mathrm{Tr}\,y\,T^N/\mathrm{Tr}\,T^N.$$

However, we know that, for large N, if λ_0 is the largest eigenvalue of the real symmetric operator T, then $(T/\lambda_0)^N (y, y') \to \phi_0(y)\,\phi_0(y')$, where ϕ_0 is the normalized eigenfunction belonging to λ_0, and so

$$E(y) = \int y\, \phi_0(y)^2\, dy.$$

Finding $\phi_0(y)$ depends of course on the precise potential $U(y)$ that is used, but the two general possibilities are easily seen. If λ_0 is part of the discrete spectrum, then $\phi_0(y)$ is localized and a finite $E(y)$ results; if there is no discrete spectrum, $\phi_0(y)$ can be normalized only if the domain of y is restricted to being finite, and then $E(y)$ will diverge as the size of the domain increases. There will thus be a phase transition to unbounded transverse motion if the discrete spectrum disappears at some value of β. Clearly,

$$\lambda_0 = e^{-s_0}(2\pi/\beta K)^{1/2}$$

where

$$-\frac{1}{2\beta K}\, \phi_0''(y) + \beta \hat{U}(y)\, \phi_0(y) = s_0\, \phi_0(y),$$

where s_0 is the *lowest* eigenvalue of the accompanying operator and $\hat{U}$ is the modified potential, as above.

Now an even one-dimensional potential (the crucial point is that $\hat{U}(\infty) = \hat{U}(-\infty)$) with a trough below its lowest asymptotic value will always have a localized discrete eigenvalue state. However, $\hat{U}$ is not of this form, and so for sufficiently high temperature – low β – there will be no bound state, leading to a γ-probability distribution stretching out uniformly to infinity. A quick estimate of the temperature at which rapid delocalization in this sense occurs is given by the standard JWKB method of solving the Schrödinger equation satisfied by $\phi_0(y)$ and gives precisely the same result as a "semiclassical" hand-waving estimate, which is as follows. In the usual Schrödinger equation of quantum physics, say for unit mass, the parameter βK is equated with $1/\hbar^2$, where $\hbar$ denotes $(1/2\pi)$ times Planck's constant, and then, not distinguishing between $\hat{U}$ and U, the classical mechanics being analyzed has the energy $E = \frac{1}{2} p^2 + \beta U(y)$. Each discrete eigenstate of energy E covers a volume $2\pi\hbar \to 2\pi/\sqrt{\beta K}$ in (y, p) space, a shell whose midsurface is precisely $\frac{1}{2} p^2 + \beta U(y) = E$. Hence the number of discrete eigenstates up to the energy value E is given by

$$N(E) - \frac{1}{2} = \iint_{\frac{1}{2}p^2 + \beta U(y) \leq E} dy\, dp/(2\pi/\sqrt{\beta K})$$

$$= \frac{1}{\pi}\, \sqrt{\beta K} \int \{2[E - \beta U(y)]\}^{1/2}\, dy,$$

integrated over the range in which the square root is real, and so the number of states up to the continuum, which starts at $E = U(\infty)$, dips below 1 when

$$\frac{1}{2}\pi/\sqrt{\beta K} = \iint 2\beta[U(\infty) - U(y)]^{1/2}\, dy,$$

or

$$\frac{1}{\beta} = (\sqrt{8k}/\pi) \int [U(\infty) - U(y)]^{1/2}\, dy.$$

In particular, for the (Morse) potential of the Peyrard–Bishop model,

$$\frac{1}{\beta} = \sqrt{8k}/\pi \int_{-\frac{1}{a}\ln 2}^{\infty} D^{1/2}(2e^{-ay} - e^{-2ay})^{1/2}\, dy$$

$$= (4/\pi a)\sqrt{KD}$$

5.2. Dynamics

The fact that the original two-strand model "evaporated" to separation in thermal equilibrium [Zh97] is a consequence of the energy fluctuations available in a thermal ensemble defined as being supplied by a heat bath, e.g., the aqueous environment. A more relevant question might be this: Suppose that the temperature is low enough that unseparated pairs of strands are at least metastable, and localized energy is supplied, e.g., by RNA polymerase; what then will be the time development of the resulting strand separation profile? To start with, let us anchor the pair of strands in its potential minimum $\{y_n = \bar{y}\}$ and look at the small-amplitude motions in the vicinity of this state. For this purpose, then

$$H_y \rightarrow \sum \frac{m}{2}\dot{y}_n^2 + \sum U(\bar{y}) + \frac{1}{2}\sum U'(\bar{y})(y_n - \bar{y})^2 + \frac{K}{2}\sum(y_n - y_{n-1})^2,$$

so the equations of motion for $\{\zeta_n = y_n - y_0\}$ take the form $\ddot{\zeta}_n - \omega_0^2(\zeta_{n+1} - 2\zeta_n + \zeta_{n-1}) + \omega_1^2 \zeta_n = 0$ (where $\omega_0^2 = K/m$ and $\omega_1^2 = U''(\bar{y})/m$). Rather than solve this easily solvable linear equation, let us observe that, on summing over n,

$$\frac{d^2}{dt^2}\sum \zeta_n + \omega_1^2 \sum \zeta_n = 0,$$

so if, e.g., we add pure kinetic energy at time 0, with $\sum \zeta_n(0) = 0$, then

$$\sum \zeta_n(t) = A \sin \omega_1 t.$$

On the other hand, multiplying by n^2 and summing, we obtain

$$\frac{d^2}{dt^2} \sum n^2 \zeta_n + \omega_0^2 \sum n^2 \zeta_n = 2\omega_1^2 \sum \zeta_n,$$

with the solution

$$\sum n^2 \zeta_n(t) = -A \frac{\omega_1^2}{\omega_0} t \cos \omega_1 t.$$

In other words, not only does an initial localized excitation oscillate as expected, but it spreads spatially as well.

The above linear analysis need not be valid beyond a short time. At longer time, the spreading of any initial distribution is very much affected by the omitted nonlinear terms. There are many ways of seeing this, but perhaps the simplest is that of equivalent linearization (Krylov and Bogoliubov, 1947) an intelligent modification of the familiar variation-of-constants approach. Suppose that the potential minimum occurs at $\bar{y} = 0$, as it does in the Peyrard–Bishop model. The basic solution of the linearly truncated dynamics is of course exponential, in both the variables n and t : $y_n = e^{i[n\theta - \omega(\theta)t]}$, where substitution into the linearized equation shows that

$$\omega(\theta)^2 = \omega_1^2 + 2\omega_0^2(1 - \cos\theta).$$

We now create a time-dependent envelope:

$$y_n(t) = \frac{1}{2} \left\{ F_n(t) e^{i[n\theta - \omega(\theta)t]} + F_n^*(t) e^{-i[n\theta - \omega(\theta)t]} \right\}$$

and ask when this can satisfy the full equation.

A neat procedure is to work with the Lagrangian rather than the Hamiltonian:

$$L_y = \sum \frac{m}{2} \dot{y}_n(t)^2 - \frac{K}{2} \sum [y_n(t) - y_{n-1}(t)]^2 - \sum U[y_n(t)],$$

and, because we are interested in the slowly varying envelope $F_n(t)$, we average the Lagrangian over a few cycles of the basic oscillation. Thus by setting

$$\phi = n\theta - \omega(\theta) t$$

and then using $\langle e^{iK\phi} \rangle = 0$ for any integer $K \neq 0$, we have

$$\left\langle \sum \dot{y}_n^2 \right\rangle = \frac{1}{4} \sum \langle [(\dot{F}_n - i\omega F_n)e^{i\phi} + (\dot{F}_n^* + i\omega F_n^*)e^{-i\phi}]^2 \rangle$$

$$= \frac{1}{2} \sum (\dot{F}_n - i\omega F_n)(\dot{F}_n^* + i\omega F_n^*)$$

and similarly

$$\left\langle \sum (y_n - y_{n-1})^2 \right\rangle = \frac{1}{4} \sum \langle [(F_n - F_{n-1}\, e^{-i\theta})e^{i\phi} + (F_n^* - F_{n-1}^*\, e^{i\theta})e^{-i\phi}]^2 \rangle$$

$$= \frac{1}{2} \sum (F_n - F_{n-1}\, e^{-i\theta})(F_n^* - F_{n-1}^*\, e^{i\theta}),$$

whereas for $U(y) = (e^{-ay} - 1)^2$, we use

$$y_n = |F_n| \cos \left(\theta + \frac{1}{2i} \ln F_n/F_n^* \right)$$

to obtain (I_0 and I_1 are the usual modified Bessel functions)

$$\left\langle \sum U(y_n) \right\rangle = D \sum \langle e^{-2ay_n} - 2e^{-ay_n} + 1 \rangle$$

$$= D \sum [I_0(2a|F_n|) - 2I_0(a|F_n|) + 1].$$

Hence

$$\langle L_y \rangle = \frac{m}{4} \sum (\dot{F}_n - i\omega F_n)(\dot{F}_n^* + i\omega F_n^*)$$

$$- \frac{K}{4} \sum (F_n - F_{n-1}\, e^{-i\theta})(F_n^* - F_{n-1}^*\, e^{i\theta})$$

$$- D \sum [I_0(2a|F_n|) - 2I_0(a|F_n|) + 1],$$

yielding the equations of motion $(d/dt)(\partial \langle L_n \rangle / \partial \dot{F}_n^*) = (\partial \langle L_n \rangle / \partial F_n^*)$, or

$$\ddot{F}_n - 2i\omega\, \dot{F}_n - \omega^2 F_n = \omega_0^2 (F_{n+1}\, e^{i\theta} - 2F_n + F_{n-1}\, e^{-i\theta})$$

$$- \frac{4aD}{m} F_n \frac{1}{|F_n|} [I_1(2a|F_n|) - I_1(a|F_n|)].$$

Because we know that $F_n = $ const is a solution at small F_n, we subtract $-\omega^2 F_n = 2\omega_0^2(\cos\theta - 1)F_n - [(4aD)/m]\frac{a}{2}F_n$, giving us

$$\ddot{F}_n - 2i\omega\dot{F}_n = \omega_0^2 \cos\theta\, (F_{n+1} - 2F_n + F_{n-1}) + i\omega^2 \sin\theta\, (F_{n+1} - F_{n-1})$$

$$- \frac{4aD}{m} F_n \frac{1}{|F_n|} \left\{ I_1(2a|F_n|) - \left[I_1(a|F_n|) - \frac{1}{2}a|F_n| \right] \right\}.$$

If the envelope changes slowly in space on the scale of base–base separation, we can replace this with

$$\ddot{F}_n - \omega_0^2 \cos\theta\, F_n'' = 2i\omega\, \dot{F}_n + 2i\omega_0^2 \sin\theta\, F_n'$$

$$- \frac{4aD}{m} F_n \frac{1}{|F_n|} \left[I_1(2a|F_n|) - I_1(a|F_n|) - \frac{1}{2}a|F_n| \right],$$

in which form some qualitative aspects are apparent. For example, imagine a frequency-shifted traveling envelope (Remoissenet, 1986)

$$F_n(t) = e^{i\delta t} F(n - vt),$$

so that

$$\left(\omega_0^2 \cos\theta - v^2\right) F'' + 2i\left[\omega_0^2 \sin\theta - (\omega - \delta)v\right] F'$$
$$= \left[2\omega_1^2 + f\left(a|F|\right) - \delta^2\right] F,$$

where

$$\omega_1^2 = 2a^2 D/m, \quad f(x) = \frac{1}{x}\left[I_1(2x) - I_1(x) - \frac{1}{2}x\right].$$

This will have a real solution if

$$v = \omega_0^2 \sin\theta/(\omega - \delta)$$

[note that because $\omega^2 = \omega_1^2 + 2\omega_0^2\left(1 - \cos\theta\right)$, then $\omega\left(d\omega/d\theta\right) = \omega_0^2 \sin\theta$, so $v_g = \omega_0^2 \sin\theta/\omega$ is the group velocity of a wave packet), and then will take the form

$$MF'' = -V'(F),$$

with the double-hill potential $V(F)$, as shown. This, we can see, means that at finite amplitude we have a solution (dashed line in figure) with a moving "kink" if $M > 0$, but otherwise (and for $M < 0$) only a moving wave train.

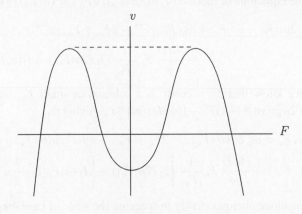

5.3. Effect of Heterogeneity

Of course, the essence of DNA lies in its heterogeneity, and we typically might expect that varying the chain–chain interactions, e.g., having a random set

$\{U_n(y_n)\}$ rather than a common one $U(y_n)$, would impart a detailed structure to the dynamics, as well as to the thermally equilibrated state. Numerical work has shown that this is not the case. However, a hint as to what might be going on is supplied by the observation that if the stiffness K in the basic model is given a y_n dependence, e.g.,

$$K \rightarrow K_1 + (K_2 - K_1) e^{-y_n/b},$$

thereby accentuating the effect of any fluctuations in the $\{y_n\}$, the homogeneous "melting" transition appears to sharpen to first order (Peyrard, 1988). This suggests that it is the variation in stiffness that is responsible for spatially preferred domains of separation. Simulations with the randomness applied to $K(y)$ instead (Cule and Hwa, 1997) have also verified this supposition. No corresponding analytic work exists.

Assignment 8

1. Examine the statistics of the traditional score $W = \sum f_{\alpha j} x_j(\alpha)$ of Subsection 3.3.3.
2. Discuss the relevance of the Arnold–Kolmogorov theorem on the recursive representation of many-variable functions to design of neural networks.
3. How is the Peyrard-Bishop analysis of H-bond breaking modified by inhomogeneity in the twin strand?

Bibliography

(A very incomplete list, restricted to those I happened to find useful)

Abremski, A., Sirotkin, K. M., and Lapedes, A., "Application of neural networks and information theory to the identification of *E. coli* transcriptional promoters," Los Alamos Report LA-UR-91-729 (Los Alamos National Laboratory, Los Alamos, NM, 1991).

Alberts, B., Bray, D., Lewis, J., Raff, M., Roberts, K., and Watson, J. D., *Molecular Biology of the Cell* (Garland, New York, 1989).

Allegrini, P., Buiatti, M., Grigolini, P., and West, B. J., "Non-Gaussian statistics of anomalous diffusion: the DNA sequences of prokaryotes," *Phys. Rev. E* **58**, 3640–3648 (1998).

Almirantis, Y. and Provata, A., "Long and short-range correlations in genome organization," *J. Stat. Phys.* **97**, 233–262 (1999).

Apostolico, A., Book, M. E., Lonardi, S., and Xu, X., "Efficient detection of unusual words," *J. Comput. Biol.* **7**, 71–94 (2000).

Arratia, R., Lander, E. S., Tavare, S., and Waterman, M. S., "Genomic mapping by anchoring random clones: a mathematical analysis," *Genomics* **11**, 806–827 (1991).

Arratia, R., Goldstein, G., and Gordon, L., "Poisson approximation and the Chen–Stein Method," *Stat. Sci.* **5**, 403–434 (1990).

Baldi, P. and Brunak, S., *Bioinformatics* (MIT Press, Cambridge, MA, 1998).

Balding, D. J. and Torney, D. C., "The design of pooling experiments for screening a clone map," *J. Fungal Gen. Biol.* **21**, 302–307 (1997).

Beale, R. and Jackson, T., *Neural Computing* (Hilger, London, 1990).

Bell, G. I., "Evolution of simple sequence repeats," *Comput. Chem.* (1994).

Benner, S. A., Cohen, M. A., and Garnet, G. H., "Empirical and structural models for insertions and deletions in the divergent evolution of proteins," *J. Mol. Biol.* **229**, 1065–1082 (1993).

Bernaola-Galván, P., Grosse, I., Carpena, P., Oliver, J. L., Román-Roldán, R., and Stanley, H. E., "Finding borders between coding and noncoding DNA regions by entropic segmentation method," *Phys. Rev. Lett.* **85**, 1342–1345 (2000).

Bernardi, G., "The isochore organization of the human genome," *Ann. Rev. Genet.* **23**, 637–661 (1989).

129

Berthelsen, C. L., Glazier, J. A., and Skolnick, M. H., "Global fractal dimension of human DNA sequences treated as pseudorandom walks," *Phys. Rev. A* **45**, 8902–8913 (1992).

Bernaola-Galván, P., Román-Roldán, R., and Oliver, J. L., "Compositional segmentation and long-range fractal correlations in DNA sequences," *Phys. Rev. E* **53**, 5181–5189 (1996).

Bishop, D. T., et al., "Number of polymorphic DNA clones required to map human genome," in Weir (1983).

Botstein, D., White, R. L., Skolnick, M., and Davis, R. W., "Construction of a genetic linkage map in man using restriction fragment length polymorphisms," *Am. J. Hum. Genet.* **32**, 314–331 (1980).

Bucher, P., "Weight matrix description of four eukaryotic RNA polymerase II promoter elements," *J. Mol. Biol.* **212**, 563–578 (1990).

Chvatal, V. and Sankoff, D., "Longest common subsequence of two random sequences," *J. Appl. Probab.* **12**, 306–315 (1975).

Chvatal, V. and Sankoff, D., "An upper-bound technique for lengths of common subsequences," in Sankoff and Kruskal (1983), pp. 353–358.

Chen, L. H. Y., "Poisson approximation for dependent trails," *Ann. Prob.* **3**, 534–545 (1975).

Churchill, G. A., "Stochastic models for heterogeneous DNA sequences," *Bull. Math. Biol.* **51**, 79–94 (1989).

Clarke, L. and Carbon, J., "A colony bank containing synthetic Col E1 hybrid plasmids representative of the entire E coli gene," *Cell* **9**, 91–101 (1976).

Cooper, N. G., "The human genome project," *Los Alamos Sci.* **20**, XX–XX (1992).

Cox, D. R., Burmeister, M., Price, E. R., Kim, S., and Myers, R. M., "Radiation hybrid mapping," *Science* **250**, 245–250 (1990).

Cramer, H., *Mathematical Methods of Statistics* (Princeton, Univ. Press, Princeton, NJ, 1946), Chap. 30.

Cule, D. and Hwa, T., "Denaturation of heterogeneous DNA," *Phys. Rev. Lett.* **79**, 2375–2378 (1997).

Czabarlia, E., Konjevod, G., Marathe, M. V., Percus, A. G., and Torney, D. C., "Algorithms for optimizing production DNA sequencing," in *Proceedings of the Annual ACM-SIAM Symposium (SODA)* (Society for Industrial and Applied Mathematics, Philadelphia, 2000), pp. 399–408.

Deken, J., "Probabilistic behavior of longest-common-subsequence length," in Sankoff and Kruskal (1983), pp. 359–362.

Doritt, R. L., Schoenback, L., and Gilbert, W., "How big is the Universe of exons?," *Science* **250**, 1377–1382 (1990).

Drasdo, D., Hwa, T., and Lassig, M., "Scaling laws and similarity detection in sequence alignment with gaps," *J. Comput. Biol.* **7**, 115–141 (2000).

Durbin, R., Eddy, S., Krogh, A., and Mitchinson, G., *Biological Sequence Analysis* (Cambridge Univ. Press, New York, 1998).

Dwyer, D. S., "Assembly of exon from unitary transposable genetic elements: implications for the evolution of protein–protein interactions," *J. Theor. Biol.* **194**, No. 1, 11–27 (1998).

Elton, R. A., "Theoretical Models for Heterogeneity of Base Composition in DNA," *J. Theor. Biol.* **45**, 533–553 (1974).

Farber, D., Lapedes, A. and Sirotkin, K. M., "Determination of eukaryotic coding regions," Los Alamos Report LA-UR-90-4014 (Los Alamos National Laboratory, Los Alamos, NM, 1999).

Fasulo, D., Jiang, T., Karp, R. M., Settergren, R. J. and Thayer, E., "An algorithmic approach to multiple complete digest mapping," *J. Comput. Biol.* **6**, 187–207 (1999).

Feller, W., *Probability Theory and its Applications* (Wiley, New York, 1950), Vol. 1.

Fickett, J. W., Torney, D. C., and Wolf, D. R., "Base compositional structure of genomes," *Genomics* **13**, 1056–1064 (1992).

Fickett, J. W. and Tung, C. S., "Assessment of protein coding measures," *Nucl. Acids Res.* **20**, 6441–6450 (1992).

Frechet, M., in *"Probabilités Associees a un Systeme d'Évenements Compatibles et Dépendent,"* (Hermann et Cie, Paris 1940).

Gilbert, W., "The exon theory of genes," *Cold Spring Harbor Symp. Quant. Biol.* **L11**, 901–905 (1997).

Gindikin, S., *Mathematical Methods of Analysis of DNA Sequences*, Vol. 8 of DIMACS Series (American Mathematical Society, Providence, RI, 1992).

Goldstein, L., "Poisson approximation and DNA sequence matching," *Commun. Stat. Theory Meth.* **19**, 4167 (1990).

Goss, S. J. and Harris, H., "New method for mapping genes in human chromosomes," *Nature (London)* **255**, 680–684 (1975).

Guibas, I. and Odlyzko, A., "Long repetition patterns in random sequences," *Z. Wahrscheinlichkeitstheor. Verwandte Geb.* **53**, 241–262 (1980).

Gumbel, J., "Statistics of extremes" (Columbia University Press, 1958).

Gusfield, D., *Algorithms on Trees, Strings and Sequences* (Cambridge Univ. Press, New York, 1997).

Hao, B., Xie, H., Yu, Z., and Chen, G., "A combinatorial problem related to avoided and under-represented strings in bacterial complete genomes," *Ann. Comb.* (2000).

Hausner, M. and Schwartz, J. T., "Lie groups, Lie Algeras" (Gordon and Breach, 1968).

Hayashi, C., "On the prediction of phenomena from qualitative data," *Ann. Inst. Stat. Math.* **3**, 69–96 (1952).

Herzel, H., Ebeling, W., and Schmitt, A. O., "Entropics of biosequences: the role of repeats," *Phys. Rev. E* **50**, 5061–5071 (1994).

Herzel, H. and Grosse, I., "Correlations in DNA sequences: the role of protein coding segments," *Phys. Rev. E* **55**, 800–810 (1997).

Hwa, T. and Lassig, M., "Similarity detection and localization," *Phys. Rev. Lett.* **76**, 2591–2594 (1996).

Iida, Y., "Nucleotide sequences of DNA and their structural analysis," *Bull. Chem. Soc. Jpn.* **60**, 2977–2981 (1987).

Iida, Y., "Categorical discriminant analysis of 3'-splice site signals of mRNA precursors in higher eukaryote genes," *J. Theor. Biol.* **135**, 109–118 (1988).

Karlin, S. and Brendel, V., "Patchiness and correlations in DNA sequences," *Science* **259**, 677–680 (1993).

Karlin, S., Ost, F., and Blaisdell, B. E., in Waterman (1989a), Chap. 6.

Karlin, S. and Macken, C., "Some statistical problems in the assessment of inhomogeneities of DNA sequence data," *J. Am. Stat. Assoc.* **86**, 27–35 (1991).

Karp, R. M. and Shamir, R., "Algorithms for optical mapping," *J. Comput. Biol.* **7**, 303–316 (2000).

Karlin, S. and Ost, F., "Maximum length of common words among random letter sequences," *Ann. Prob.* **16**, 533–563 (1988).

Katz, R. W., "On some criteria for estimating the order of a Markov chain," *Technometrics* **23**, 243–249 (1981).

Kim, S. and Segre, A. M., "AMASS: a structural pattern matching approach to shotgun sequence assembly," *J. Comput. Biol.* **6**, 163–186 (1999).

Kryloff, N. and Bogoliuboff, N., *Introduction to Non-Linear Mechanics* (Princeton Univ. Press, Princeton, NJ, 1947).

Lander, B. S. and Waterman, M. S., "Genomic mapping by fingerprinting random clones: a mathematical analysis," *Genomics* **2**, 231–239 (1988).

Lander, B. S., "Analysis with restriction enzymes," in Waterman (1989a), Chap. 2.

Lange, K. and Boehnke, M., "How many polymorphic genes will it take to span the human genome?," *Am. J. Hum. Genet.* **34**, 842–845 (1982).

Lathrop, R. H. and Smith, T. G., "Global optimum protein threading with gapped alignment and empirical pair score functions," *J. Mol. Biol.* **235**, 641–665 (1996).

Li, W., "Mutual information function versus correlation functions," *J. Stat. Phys.* **60**, 823–827 (1990).

Li, W., "The study of correlation structures of DNA sequences: a critical review," *Comput. Chem.* **21**, 257–272 (1997).

Li, W., "Statistical properties of open reading frames in complete genome sequences," *Comput. Chem.* (1999).

Loewenstern, D. and Ylanilos, D. N., "Significantly lower entropy estimate for natural DNA sequence," *J. Comput. Biol.* **6**, 125–142 (1999).

Li, W. and Kaneko, K., "Long range correlation and partial $1/f^\alpha$ spectrum in a noncoding DNA sequence," *Europhys. Lett.* **17**, 655–660 (1992).

Li, W., Marr, T. G., and Kaneko, K., "Understanding long-range correlations in DNA sequences," *Physica D* (1994).

Lindenmeyer, H., *J. Theor. Biol.* **18**, 280 (1968).

Lu, X., Sun, Z., Chen, H., and Li, Y., "Characterizing self-similarity on bacteria DNA sequences," *Phys. Rev. E* **58**, 3578–3584 (1998).

Maddox, J., *Nature* (London) **339**, (1989).

Mayraz, G. and Shamir, R., "Construction of physical maps from oligonucleotide fingerprints data," *J. Comput. Biol.* **6**, 237–252 (1999).

Meyer, Y., *Wavelets and Operators* (Cambridge Univ. Press, New York, 1992).

Miura, R. M., Some Mathematical Questions in Biology, Vol. 17 of DNA Sequence Analysis Series (American Mathematical Society, Providence, RI, 1986).

Miyazawa, S. and Jernigan, R. L., "Estimation of effective interresidue contact energies from protein crystal structure," *Macromolecules* **18**, 534–552 (1985).

Mohanty, A. K. and Narayana Rao, A. V. S. S., "Factorial moments analyses show a characteristic length scale in DNA sequence," *Phys. Rev. Lett.* **84**, 1832–1835 (2000).

Mount, S. M., *Nucl. Acids Res.* **10**, 459 (1982).

Mott, R. F., Kirkwood, T. B. L., and Kurnow, R. N., "An accurate approximation to the distribution of the length of the longest matching word between two random sequences," *Bull. Math. Biol.* **52**, 773–784 (1990).

Mott, R. and Tribe, R., "Approximate statistics of gapped alignments," *J. Comput. Biol.* **6**, 91–112 (1999).

Nathans, D. and Smith, H. O., "Restriction endonucleases," *Ann. Rev. Biochem.* **44**, 273–298 (1975).

Needleman, S. B. and Wunsch, C. D., "A general method applicable to the search for similarities in the amino acid sequence of two proteins," *J. Mol. Biol.* **48**, 443–453 (1970).

Otto, G., New York University, New York, NY 10012 (personal communication, 1992).

Peng, C. K., Buldyrev, S. V., Goldberger, A. L., Havlin, S., Sciortino, F., Simons, M., and Stanley, H. E., "Long-range correlations in nucleotide sequences," *Nature (London)* **356**, 168–170 (1992).

Peng, C. K., Buldyrev, S. V., Havlin, S., Simons, M., Stanley, H. E., and Goldberger, A. L., "Mosaic organization of DNA nucleotides," *Phys. Rev. E* **49**, 1685–1689 (1994).

Percus, O. E. and Percus, J. K., "String matching for the novice," *Am. Math. Monthly* **102**, 944–947 (1994).

Percus, J. K., *Combinatorial Methods* (Springer-Verlag, New York, 1971).

Percus, O. E., Percus, J. K., Bruno, and Torney, D. C., "Asymptotics of pooling design performance," *J. Appl. Prob.* **36**, 951–964 (1999).

Percus, O. E. and Percus, J. K., "Island length distribution in genome sequencing," *J. Math. Biol.* **39**, 244–268 (1999).

Percus, J. K., Percus, O. E., and Perelson, A. S., "Probability of self–nonself discrimination," *NATO ASI Series* **H16**, 63–70 (1992).

Peyrard, M., "Nonlinear energy localization in biomolecules" in *Biological Physics*, Frauenfelder, H., Hummer, G., Garcia, R., eds. (American Institute of Physics, New York, 1998), p. 147–162.

Peyrard, M. and Bishop, A. R., "Statistical mechanics of a nonlinear model for DNA denaturation," *Phys. Rev. Lett.* **62**, 2755–2758 (1989).

Port, E., Sun, F., Martin, D., and Waterman, M. S., "Genomic mapping by end-characterized random clones: a mathematical analysis," *Genomics* **26**, 84–100 (1995).

Poustka, A., et al., "Molecular approaches to mammalian genetics," *Cold Spring Harbor Symp. Quant. Biol.* **51**, 131–139 (1986).

Raftery, A. E., "A model for high order Markov chains," *J. R. Stat. Soc. Ser. B* **47**, 528 (1985).

Ramensky, V. E., Makeev V. Ju., Roytberg, M. A., and Tuasuyan, V. G., "DNA segmentation through the Bayesian approach," *J. Comput. Biol.* **7**, 218–231 (2000).

Remoissenet, M., "Low-amplitude breather and envelope solitons on quasi-one-dimensional physical models," *Phys. Rev. E* **38**, 2386–2392 (1986).

Robbins, H., "On the measure of a random set," *Ann. Math. Stat.* **15**, 70–74 (1994).

Román-Roldán, R., Bernaola-Galván, P., and Oliver, J. L., "Sequence compositional complexity of DNA through an entropic segmentation method," *Phys. Rev. Lett.* **80**, 1344–1347 (1998).

Saiki, R. K., Gelfand, D. H., Stoffel, S., Scharf, S. J., Hiaguchi, R., Horn, G. T., Mullis, K. B., Ehrlich, H. A., "Primer-directed enzymatic amplification of DNA with a thermostable DNA polymerase," *Science* **239**, 487–491 (1988).

Sanger, F., et al., "DNA sequencing with chain terminating inhibitors," *Proc. Natl. Acad. Soc. USA* **74**, 5463–5467 (1977).

Sankoff, D. and Kruskal, eds., *Time Warps, String Edits, and Macromolecules* (Addison-Wesley, Reading, MA, 1983).

Sankoff, D. and Mainville, S., "Common subsequences and monotone subsequences," in Sankoff and Kruskal (1983).

Schbath, S., Bossard, N., and Tavare, S., "The effect of nonhomogeneous clone length distribution on the progress of an STS mapping project," *J. Comput. Biol.* **7**, 47–57 (2000).

Schaffer, H. E., "Determination of DNA fragment size from gel electrophoresis mobility," in Weir (1983), pp. 1–14.

Schaffer, H. E. and Sederoff, R. R., "Improved estimation of DNA fragment lengths from agarose gels," *Anal. Biochem.* **115**, 113–123 (1981).

Schmitt, W. and Waterman, M. S., "Multiple solutions of DNA restriction mapping problem," *Adv. Appl. Math.* **12**, 412–427 (1991).

Schwartz, D. C., Li, X., Hernandez, L. I., Ramnarain, S. P., Huff, E. J., and Wang, Y. K., "Ordered restriction maps of saccharomyces chromosomes constructed by optical mapping," *Science* **262**, 110–114 (1993).

Selim, S. Z. and Alsultan, K., "A simulated annealing algorithm for the clustering problem," *Pattern Recogn.* **24**, 1003–1008 (1991).

Sellers, P. H., "The mathematical meaning of pattern similarities in genetic structures," in *Mathematics and Computers in Biomedical Applications*, Eisenfeld and de Lisi, eds., 23–27 (Elsevier, New York, 1985), Chap. 23.

Sellers, P. H., "Pattern recognition in DNA," in American Mathematical Society (1986), Chap. 19.

Shapiro, M. D. and Senapathy, P., "RNA splice junctions of different classes of eukaryotes," *Nucl. Acids Res.* **15**, 7155–7174 (1987).

Slomin, D., et al., "Building human genome maps with radiation hybrids," *Proc. Recomb.* **97**, 277–286 (1997).

Stolorz, Lapedes, and Xia, "Predicting protein secondary structure using neural net and statistical methods," *J. Mol. Biol.* (1992).

Stoltzfus, A., Spencer, D. F., Zuker, M., Logsdon, J. M., and Doolittle, W. F., "Testing the exon theory of genes: the evidence from protein structure," *Science* **265**, 202–207 (1994).

Stormo, G. D. and Hartzell, G. W., "Identifying protein-building sites from unaligned DNA fragments," *Proc. Natl. Acad. Sci. USA* **86**, 1183–1187 (1989).

Tavare, S. and Giddings, B. W., "Some statistical aspects of the primary structure of nucleotide sequence," in Waterman (1989a), Chap. 5.

Techera, M., Daeman, L. L., and Prohovsky, E. W., "Nonlinear model of the DNA molecule," *Phys. Rev. A* **40**, 6636–6642 (1989).

Tong, H., "Determination of the order of a Markov chain by Akaike's information criterion," *J. Appl. Probab.* **12**, 488–497 (1975).

Torney, D. C., Burks, C., Davison, D., and Sirotkin, K. M., "Computation of d^2: a measure of sequence dissimilarity," in *Computers and DNA*, G. Bell and T. Marr, eds. (Addison-Wesley, Reading, MA, 1990), pp. 109–125.

Torney, D. C., "Mapping using unique sequences," *J. Mol. Biol.* **217**, 259–264 (1991).

Trifonov, E. N., "Nucleotide sequences as a language: morphological classes of words," in *Classification and Related Methods of Data Analysis*, H. H. Book, ed. (Elsevier, New York, 1988), pp. 57–64.

Tsonis, A. A., Kuman, P., Elsner, J. B., and Tsonis, P. A., "Wavelet analysis of DNA sequence," *Phys. Rev. E* **53**, 1828–1834 (1996).

Uspensky, J. V., *Introduction to Mathematical Probability* (McGraw-Hill, New York, 1937).

Voss, R. F., "Evolution of long-range fractal correlations and $1/f$ noise in DNA base sequences," *Phys. Rev. Lett.* **68**, 3805–3808 (1992).

Waterman, M. S., "Frequencies of restriction sites," *Nucl. Acids Res.* **11**, 8951–8956 (1983).

Waterman, M. S., "Probability distribution for DNA sequence comparisons," *Lectures Math Life Sci.* **17**, 29–56, American Math Soc., Providence (1986).

Waterman M. S., ed., *Mathematical Methods for DNA Sequences* (CRC, Boca Raton, FL, 1989a).

Waterman, M. S., *Consensus Patterns in Sequences*, in Warerman (1989a), Chap. 4.

Waterman, M. S., *Consensus Methods for Folding Single Stranded Nucleic Acids*, in Waterman (1989a), Chap. 8.

Waterman, M. S., *Introduction to Computational Biology* (Chapman & Hall, London, 1995).

Waterman, M. S., Gordon, L., and Arratia, R., "Phase transitions in sequence matches and nucleic acid structure," *Proc. Natl. Acad. Sci. USA* **84**, 1239–1242 (1987).

Weir, B. S., ed., *Statistical Analysis of DNA Sequence Data* (Marcel Dekker, New York, 1983).

Weir, B. S., ed., *Genetic Data Analysis* (Sinauer, Sunderland, Mass., 1990).

Zhang, I. I. Q. and Marr, T. G., Cold Spring Harbor, Long Island, NY (personal communication, 1992).

Zhang, M. Q. and Marr, T. G., "Genome mapping by random anchoring: a discrete theoretical analysis," *J. Stat. Phys.* **73**, 611–623 (1994a).

Zhang, M. Q. and Marr, T. G., "Statistical analysis of the fission yeast genomic DNA sequence and gene recognition," *Nucl. Acid Res.* **22**, 1750–1759 (1994b).

Zhang, M. Q. and Marr, T. G., "Molecular sequence alignment seen as random path analysis," *J. Theor. Biol.* **174**, 119–130 (1995).

Zhang, Y., Zheng, W., Liu, J., and Chen, Y. Z., "Theory of DNA melting based on the Peyrard-Bishop Model," *Phys. Rev. E* **56**, 7100–7115 (1997).

Index